HOW TERRORIST GROUPS END

Lessons for Countering al Qa'ida

Seth G. Jones • Martin C. Libicki

This research in the public interest was supported by RAND, using discretionary funds made possible by the generosity of RAND's donors, the fees earned on client-funded research, and independent research and development (IR&D) funds provided by the Department of Defense.

Library of Congress Cataloging-in-Publication Data

Jones, Seth G., 1972–
How terrorist groups end : lessons for countering Al Qa'ida / Seth G. Jones, Martin C. Libicki.
p. cm.
Includes bibliographical references.
ISBN 978-0-8330-4465-5 (pbk. : alk. paper)
1. Terrorism. 2. Terrorism—Prevention—International cooperation. 3. Intelligence service. 4. Problem-oriented policing. 5. Qaida (Organization) I. Libicki, Martin C. II. Title.

HV6431.J65 2008
363.325'16—dc22

2008025194

The RAND Corporation is a nonprofit research organization providing objective analysis and effective solutions that address the challenges facing the public and private sectors around the world. RAND's publications do not necessarily reflect the opinions of its research clients and sponsors.

Cover design by Carol Earnest

Published 2008 by the RAND Corporation
1776 Main Street, P.O. Box 2138, Santa Monica, CA 90407-2138
1200 South Hayes Street, Arlington, VA 22202-5050
4570 Fifth Avenue, Suite 600, Pittsburgh, PA 15213-2665
RAND URL: http://www.rand.org/
To order RAND documents or to obtain additional information, contact
Distribution Services: Telephone: (310) 451-7002;
Fax: (310) 451-6915; Email: order@rand.org

About the Authors

Seth G. Jones is a political scientist at RAND and an adjunct professor at Georgetown University's Edmund A. Walsh School of Foreign Service and the U.S. Naval Postgraduate School. He specializes in stability operations and counterinsurgency. He is the author of *In the Graveyard of Empires: America's War in Afghanistan* (W. W. Norton, forthcoming) and *The Rise of European Security Cooperation* (Cambridge University Press, 2007). He has published articles in such journals as *International Security*, *National Interest*, *Security Studies*, *Chicago Journal of International Law*, *International Affairs*, and *Survival*, as well as such newspapers and magazines as the *New York Times*, *Newsweek*, *Financial Times*, and *International Herald Tribune*. His RAND publications include *Counterinsurgency in Afghanistan: RAND Counterinsurgency Study, Volume 4* (2008); *Establishing Law and Order after Conflict* (2005); *The UN's Role in Nation-Building: From the Congo to Iraq* (2005); and *America's Role in Nation-Building: From Germany to Iraq* (2003). He received an M.A. and Ph.D. from the University of Chicago.

Martin C. Libicki is a senior management scientist at RAND, focusing on the relationship between information technology and national security. This work is documented in commercially published books, *Conquest in Cyberspace: National Security and Information Warfare* (2007) and *Information Technology Standards: Quest for the Common Byte* (1995), as well as in numerous monographs, notably *What Is Information Warfare* (1995), *The Mesh and the Net: Speculations on Armed Conflict in a Time of Free Silicon* (1994), and *Exploring Terrorist Targeting Preferences* (2007). He was also an editor of the RAND textbook, *New Challenges, New Tools for Defense Decisionmaking* (2003). His most recent assignments were to create and analyze a database of post–World War II insurgencies, devise a strategy to maximize the use of information and information technology in countering insurgency, explore terrorists' targeting preferences, develop a post–September 11 information technology strategy for U.S. Department of Justice and the Defense Advanced Research Projects Agency's Terrorist Information Awareness program, conduct an information-security analysis for the FBI, and assess CIA's R&D venture, In-Q-Tel. Other work has examined information warfare and the revolution in military affairs. Prior employment includes 12 years at the National Defense University, three years on the Navy staff as program sponsor for industrial preparedness, and three years as a policy analyst for the U.S. Government Accountability Office's Energy and Minerals Division. He received a Ph.D. from the University of California, Berkeley.

Preface

By analyzing 648 groups that existed between 1968 and 2006, this monograph examines how terrorist groups end. Its purpose is to inform U.S. counterterrorist efforts by understanding how groups have ended in the past and to assess implications for countering al Qa'ida. This monograph results from the RAND Corporation's continuing program of self-initiated independent research. Support for such research is provided, in part, by donors and by the independent research and development provisions of RAND's contracts for the operation of its U.S. Department of Defense federally funded research and development centers.

This research was conducted within the RAND National Security Research Division (NSRD) of the RAND Corporation. NSRD conducts research and analysis for the Office of the Secretary of Defense, the Joint Staff, the Unified Commands, the defense agencies, the Department of the Navy, the U.S. Intelligence Community, allied foreign governments, and foundations.

For more information on the RAND National Security Research Division, contact the Director of Operations, Nurith Berstein. She can be reached by email at Nurith_Berstein@rand.org; by phone at 703-413-1100, extension 5469; or by mail at RAND, 1200 South Hayes Street, Arlington VA 22202-5050. More information about the RAND Corporation is available at www.rand.org.

Contents

Figures

Tables

Summary

All terrorist groups eventually end. But how do they end? Answers to this question have enormous implications for counterterrorism efforts. The evidence since 1968 indicates that most groups have ended because (1) they joined the political process or (2) local police and intelligence agencies arrested or killed key members. Military force has rarely been the primary reason for the end of terrorist groups, and few groups within this time frame achieved victory. This has significant implications for dealing with al Qa'ida and suggests fundamentally rethinking post–September 11 U.S. counterterrorism strategy.

The ending of most terrorist groups requires a range of policy instruments, such as careful police and intelligence work, military force, political negotiations, and economic sanctions. Yet policymakers need to understand where to prioritize their efforts with limited resources and attention. Following an examination of 648 terrorist groups that existed between 1968 and 2006, we found that a transition to the political process is the most common way in which terrorist groups ended (43 percent). The possibility of a political solution is inversely linked to the breadth of terrorist goals. Most terrorist groups that end because of politics seek narrow policy goals. The narrower the goals of a terrorist organization, the more likely it can achieve them without violent action—and the more likely the government and terrorist group may be able to reach a negotiated settlement.

Against terrorist groups that cannot or will not make a transition to nonviolence, policing is likely to be the most effective strategy (40 percent). Police and intelligence services have better training and information to penetrate and disrupt terrorist organizations than do such institutions as the military. They are the primary arm of the government focused on internal security matters. Local police and intelligence agencies usually have a permanent presence in cities, towns, and villages; a better understanding of the threat environment in these areas; and better human intelligence.

Other reasons are less common. For example, in 10 percent of the cases, terrorist groups ended because their goals were achieved, and military force led to the end of terrorist groups in 7 percent of the cases. Militaries tended to be most effective when used against terrorist groups engaged in an insurgency in which the groups were large, well armed, and well organized. Insurgent groups have been among the most capable and lethal terrorist groups, and military force has usually been a necessary component in such cases. Against most terrorist groups, however, military force is usually too blunt an instrument. Military tools have increased in precision and lethality, especially with the growing use of precision standoff weapons and imagery to monitor terrorist movement. But even precision weapons have been of limited use against terrorist groups. The use of substantial U.S. military power against terrorist groups also runs a significant risk of turning the local population against the government by killing civilians.

Our quantitative analysis looked at groups that have ended since 1968 or are still active. It yielded several other interesting findings:

- Religious terrorist groups take longer to eliminate than other groups. Approximately 62 percent of all terrorist groups have ended since 1968, but only 32 percent of religious terrorist groups have ended.
- Religious groups rarely achieve their objectives. No religious group that has ended achieved victory since 1968.
- Size is a significant determinant of a group's fate. Big groups of more than 10,000 members have been victorious more than 25 percent of the time, while victory is rare when groups are smaller than 1,000 members.
- There is no statistical correlation between the duration of a terrorist group and ideological motivation, economic conditions,

regime type, or the breadth of terrorist goals. But there appears to be some correlation between the size of a terrorist group and duration: Larger groups tend to last longer than smaller groups.
- When a terrorist group becomes involved in an insurgency, it does not end easily. Nearly 50 percent of the time, groups ended by negotiating a settlement with the government; 25 percent of the time, they achieved victory; and 19 percent of the time, military forces defeated them.
- Terrorist groups from upper-income countries are much more likely to be left-wing or nationalist and much less likely to be motivated by religion.

Implications for al Qa'ida

What does this mean for counterterrorism efforts against al Qa'ida? After September 11, 2001, the U.S. strategy against al Qa'ida centered on the use of military force. Indeed, U.S. policymakers and key national-security documents referred to operations against al Qa'ida as *the war on terrorism*. Other instruments were also used, such as cutting off terrorist financing, providing foreign assistance, engaging in diplomacy, and sharing information with foreign governments. But military force was the primary instrument.

The evidence by 2008 suggested that the U.S. strategy was not successful in undermining al Qa'ida's capabilities. Our assessment concludes that al Qa'ida remained a strong and competent organization. Its goals were the same: uniting Muslims to fight the United States and its allies (the far enemy) and overthrowing western-friendly regimes in the Middle East (the near enemy) to establish a pan-Islamic caliphate. Al Qa'ida has been involved in more terrorist attacks since September 11, 2001, than it was during its prior history. These attacks spanned Europe, Asia, the Middle East, and Africa. Al Qa'ida's modus operandi also evolved and included a repertoire of more-sophisticated improvised explosive devices (IEDs) and a growing use of suicide attacks. Its organizational structure evolved, making it a more dangerous enemy. This included a bottom-up approach (encouraging independent action from low-level operatives) and a top-down one (issuing strategy and operations from a central hub in Pakistan).[1]

Ending the "War" on Terror

Al Qa'ida's resurgence should trigger a fundamental rethinking of U.S. counterterrorism strategy. Based on our analysis of how terrorist groups end, a political solution is not possible. Since al Qa'ida's goal remains the establishment of a pan-Islamic caliphate, there is little reason to expect that a negotiated settlement with governments in the Middle East is possible. A more effective approach would be adopting a two-front strategy.

First, policing and intelligence should be the backbone of U.S. efforts. In Europe, North America, North Africa, Asia, and the Middle East, al Qa'ida consists of a network of individuals who need to be tracked and arrested. This would require careful work abroad from such organizations as the Central Intelligence Agency (CIA) and Federal Bureau of Investigation (FBI), as well as their cooperation with foreign police and intelligence agencies. Second, military force, though not necessarily U.S. soldiers, may be a necessary instrument when al Qa'ida is involved in an insurgency. Local military forces frequently have more legitimacy to operate than the United States has, and they have a better understanding of the operating environment, even if they need to develop the capacity to deal with insurgent groups over the long run. This means a light U.S. military footprint or none at all. The U.S. military can play a critical role in building indigenous capacity but should generally resist being drawn into combat operations in Muslim societies, since its presence is likely to increase terrorist recruitment.

A key part of this strategy should include ending the notion of a war on terrorism and replacing it with such concepts as *counterterrorism*, which most governments with significant terrorist threats use. The Brit-

[1] Bruce Hoffman, "Challenges for the U.S. Special Operations Command Posed by the Global Terrorist Threat: Al Qaeda on the Run or on the March?" written testimony submitted to the U.S. House of Representatives Committee on Armed Services Subcommittee on Terrorism, Unconventional Threats, and Capabilities, February 14, 2007, p. 3.

ish government, among others, has already taken this step and abjured the phrase *war on terror*. The phrase raises public expectations—both in the United States and elsewhere—that there is a battlefield solution to the problem of terrorism. It also encourages others abroad to respond by conducting a jihad (or holy war) against the United States and elevates them to the status of holy warriors. Terrorists should be perceived and described as criminals, not holy warriors. Our analysis suggests that there is no battlefield solution to terrorism. Military force usually has the opposite effect from what is intended: It is often overused, alienates the local population by its heavy-handed nature, and provides a window of opportunity for terrorist-group recruitment. This strategy should also include rebalancing U.S. resources and attention on police and intelligence work. It also means increasing budgets at the CIA, U.S. Department of Justice, and U.S. Department of State and scaling back the U.S. Department of Defense's focus and resources on counterterrorism. U.S. special operations forces will remain critical, as will U.S. military operations to counter terrorist groups involved in insurgencies.

There is reason to be hopeful. Our analysis concludes that al Qa'ida's probability of success in actually overthrowing any government is close to zero. Out of all the religious groups that ended since 1968, none ended by achieving victory. Al Qa'ida has virtually unachievable objectives in trying to overthrow multiple regimes in the Middle East. To make matters worse, virtually all governments across Europe, North America, South America, Asia, the Middle East, and Africa consider al Qa'ida an enemy. As al Qa'ida expert Peter Bergen has noted, "Making a world of enemies is never a winning strategy."[2]

---

[2] Peter Bergen, "Al Qaeda Status," written testimony submitted to the U.S. House of Representatives Permanent Select Committee on Intelligence, April 9, 2008.

Acknowledgments

This book would not have been possible without the help of numerous individuals. Nathan Chandler provided key research support and was intimately involved in building and revising the data set. His role was fundamental, and his fingerprints are on virtually every page of the document. Daniel Byman and Lindsay Clutterbuck provided thorough reviews, which significantly improved the quality of the document. Bruce Hoffman, Gary Berntsen, and James Dobbins provided useful information and critiques. Audrey Kurth Cronin and Timothy Crawford offered excellent comments at the American Political Science Association conference in 2007. We owe a special debt of gratitude to those government officials from the United States, Afghanistan, Pakistan, Iraq, India, Canada, and European countries who took time out of their busy schedules to provide us critical information about al Qa'ida and counterterrorism efforts. In the United States, these included individuals at the National Ground Intelligence Center, Defense Intelligence Agency, Central Intelligence Agency, White House, Office of the Secretary of Defense, U.S. Agency for International Development, and U.S. Department of State. Most did not want to be identified because of the sensitivity of the information they provided, so we cannot thank them by name.

Abbreviations

ANC	African National Congress
AQI	al Qa'ida in Iraq
ARENA	Alianza Republicana Nacionalista [Nationalist Republican Alliance]
ATAG	Association Totalement Anti-Guerre [Totally Anti-War Group]
BNP	Bangladesh Nationalist Party
CI	confidence interval
CIA	Central Intelligence Agency
COPAZ	Comisión Nacional para la Consolidación de la Paz [National Commission for the Consolidation of Peace]
CPA	Coalition Provisional Authority
EOKA	Ethniki Organosis Kyprion Agoniston [National Organization of Cypriot Fighters]
ETA	Euskadi Ta Askatasuna [Basque Fatherland and Freedom]
FBI	Federal Bureau of Investigation

FLN	Front de Libération Nationale [National Liberation Front]
FLQ	Front de Libération du Québec [Liberation Front of Quebec]
FMLN	Frente Farabundo Martí para la Liberación Nacional [Farabundo Marti National Liberation Front]
GIA	Groupe Islamique Armé [Armed Islamic Group]
GNI	gross national income
GSPC	Groupe Salafiste pour la Prédication et le Combat [al-Qa'ida Organization in the Islamic Maghreb]
Hamas	Harakat al-Muqawama al-Islamiyya
IAI	Islamic Army in Iraq
ICITAP	International Criminal Investigative Training Assistance Program
IED	improvised explosive device
IRA	Irish Republican Army
ISI	Islamic State of Iraq
JAMI	Al-Jabhah al-Islamiyah al-Muqawamah al-Iraqiyah [Islamic Front for Iraqi Resistance: Salah-Al-Din Al-Ayyubi Brigades]
JeM	Jaish-e-Mohammad
JSDF	Japan Self-Defense Forces
Lehi	Lohamei Herut Israel
M-19	Movimiento 19 de Abril [April 19 Movement]
MSC	Mujahedeen Shura Council
NSRD	RAND National Security Research Division

ONUSAL	United Nations Observer Mission in El Salvador
OPEC	Organization of the Petroleum Exporting Countries
PKK	Partiya Karkerên Kurdistan [Kurdistan Workers' Party]
QPP	Quebec Provincial Police, a loose translation of Sûreté du Québec
RCMPSS	Royal Canadian Mounted Police Security Service
RENAMO	Resistencia Nacional Mozambicana [Mozambique National Resistance Movement]
RUF	Revolutionary United Front
UCLAT	l'Unité de Coordination de la Lutte Antiterroriste [Antiterrorist Coordination Unit]
USAID	U.S. Agency for International Development

CHAPTER ONE

Introduction

There has been a great deal of work on why individuals or groups resort to terrorism.[1] There has also been a growing literature on whether terrorism "works."[2] But there has been virtually no systematic analysis by policymakers or academics on how terrorism ends.[3] Methodological problems plague most of the works that have addressed the end of

---

[1] Bruce Hoffman, *Inside Terrorism*, 2nd ed., New York: Columbia University Press, 2006; Walter Reich, ed., *Origins of Terrorism: Psychologies, Ideologies, Theologies, States of Mind*, Washington, D.C.: Woodrow Wilson Center Press, 1998; Marc Sageman, *Understanding Terror Networks*, Philadelphia, Pa.: University of Pennsylvania Press, 2004; Jessica Stern, *Terror in the Name of God: Why Religious Militants Kill*, New York: HarperCollins Publishers, 2003; Martha Crenshaw, "The Causes of Terrorism," in Charles W. Kegley, ed., *International Terrorism: Characteristics, Causes, Controls*, New York: St. Martin's, 1990, pp. 92–105. On suicide terrorism, see, for example, Robert A. Pape, *Dying to Win: The Strategic Logic of Suicide Terrorism*, New York: Random House, 2005; and Mia Bloom, *Dying to Kill: The Allure of Suicide Terror*, New York: Columbia University Press, 2005, pp. 76–100.

[2] Ehud Sprinzak, "Rational Fanatics," *Foreign Policy*, No. 120, September–October 2000, pp. 66–73; Pape (2005); Alan M. Dershowitz, *Why Terrorism Works: Understanding the Threat, Responding to the Challenge*, New Haven, Conn.: Yale University Press, 2002; David A. Lake, "Rational Extremism: Understanding Terrorism in the Twenty-First Century," *Dialogue IO*, Vol. 1, No. 1, January 2002, pp. 15–29; Andrew Kydd and Barbara F. Walter, "The Strategies of Terrorism," *International Security*, Vol. 31, No. 1, Summer 2006, pp. 49–79; Robert Trager and Dessislava P. Zagorcheva, "Deterring Terrorism: It Can Be Done," *International Security*, Vol. 30, No. 3, Winter 2005–2006, pp. 87–123.

[3] There has been some work. See, for example, Audrey Kurth Cronin, *Ending Terrorism: Lessons for Defeating al-Qaeda*, Abingdon, Oxon: Routledge for International Institute for Strategic Studies, 2008; Martha Crenshaw, "Why Violence Is Rejected or Renounced: A Case Study of Oppositional Terrorism," in Thomas Gregor, ed., *A Natural History of Peace*, Nashville, Tenn.: Vanderbilt University Press, 1996, pp. 249–272; U.S. Institute of Peace, *How Terrorism Ends*, Washington, D.C., 1999.

terrorist groups. They focus on one case, such as the Irish Republican Army (IRA), or perhaps a handful of cases in which a terrorist group abandons the use of terror. As with many case studies, a small number of observations can lead to indeterminate results. They do not control for random error, which can make it extremely difficult to determine which of several alternative explanations is the most viable. The value of this study is that it looks at all terrorist groups since 1968. It is more systematic than previous analyses. In our view, the United States cannot conduct an effective long-term counterterrorism campaign against al Qa'ida or other terrorist groups without understanding how terrorist groups end.

This research seeks to fill this gap. It asks, How have terrorist groups ended in the past? What are the implications for dealing with al Qa'ida? As explained in more detail in Chapter Two, terrorist groups can end in several ways. Military or police forces defeat some; some splinter by joining existing terrorist groups or forming new ones; some calculate that they can better achieve their goals through nonviolence; and a few achieve victory. But sorting out how frequently—and how—groups end remains a puzzle.

This research has significant implications for U.S. foreign policy. In a memo to senior U.S. Department of Defense officials, for instance, former secretary of defense Donald Rumsfeld asked, "Are we winning or losing the Global War on Terror?" It continued,

> Are we capturing, killing or deterring and dissuading more terrorists every day than the madrassas and the radical clerics are recruiting, training and deploying against us? Does the US need to fashion a broad, integrated plan to stop the next generation of terrorists? The US is putting relatively little effort into a long-range plan, but we are putting a great deal of effort into trying to stop terrorists.[4]

[4] Donald H. Rumsfeld, secretary of defense, "Global War on Terrorism," memorandum to General Richard B. Myers, Paul Wolfowitz, General Peter Pace, and Douglas J. Feith, October 16, 2003.

At the core of these concerns is confusion about which counterterrorism strategies are most likely to be effective. One major reason is that there has been little systematic analysis of how terrorist groups have ended in the past.

Definitions

While there is no broadly accepted definition of *terrorism*, this monograph argues that terrorism involves the use of politically motivated violence against noncombatants to cause intimidation or fear among a target audience.[5] There are several fundamental aspects of terrorism. Terrorism has a political nature and involves the perpetration of acts designed to encourage political change. It involves the targeting of civilians. And it is restricted to organizations other than a national government. Although one could broaden the definition of terrorism to include the actions of a national government against its own or another population, adopting such a broad definition would distract attention from what policymakers would most like to know: how to combat the threat that violent substate groups pose. Further, it could also create analytic confusion. Terrorist organizations and state governments have different levels of resources, face different kinds of incentives, and are susceptible to different types of pressures. Accordingly, the determinants of their behavior are not likely to be the same and, thus, require separate theoretical investigations.[6]

A *terrorist group* is defined as a collection of individuals belonging to a nonstate entity that uses terrorism to achieve its objectives. Such an entity has at least some command and control apparatus that, no

---

[5] There are many definitions of *terrorism*. See, for example, U.S. Department of State, Office of the Coordinator for Counterterrorism, *Country Reports on Terrorism 2005*, Washington, D.C., 2006, p. 9; Hoffman (2006, pp. 1–41); Pape (2005, p. 9); and Audrey Kurth Cronin, "Behind the Curve: Globalization and International Terrorism," *International Security*, Vol. 27, No. 3, Winter 2002–2003, pp. 30–58, p. 33.

[6] Robert A. Pape, "The Strategic Logic of Suicide Terrorism," *American Political Science Review*, Vol. 97, No. 3, August 2003, pp. 343–361.

matter how loose or flexible, provides an overall organizational framework and general strategic direction.

This study also made some a priori assumptions about why groups resort to terrorism and why they may end its use.[7] First, terrorism generally has two proximate purposes: to gain supporters and to coerce opponents. Most terrorist groups seek both goals to some extent, often aiming to affect enemy calculations while mobilizing support for the terrorists' cause. In some cases, they may even try to gain an edge over rival groups. Second, terrorism is mainly utilitarian. Groups can reassess their goals and tactics in response to changes within the group, as well as changes within the larger political and social environment. Third, terrorist groups deliberately choose terrorism as one among a number of alternative means of pursuing political ends. Fourth, terrorist campaigns usually arise out of larger political movements or tendencies. Its practitioners are rarely loners, disconnected from society. Rather, they usually believe that they are acting in the interests of some larger group, such as fellow religious believers, workers, persecuted ethnic groups, or the Aryan race.[8]

Research Design

This study adopted two methodological approaches. One was to compile and analyze a data set of all terrorist groups between 1968 and 2006 (a total of 648).[9] This approach included several steps.

---

[7] On the logic of terrorism, see Jeffrey Ian Ross and Ted Robert Gurr, "Why Terrorism Subsides: A Comparative Study of Canada and the United States," *Comparative Politics*, Vol. 21, No. 4, July 1989, pp. 405–426; Dershowitz (2002); Pape (2005); Lake (2002); and Kydd and Walter (2006).

[8] Ross and Gurr (1989, p. 407).

[9] We began with the year 1968 because that is the first year the RAND-MIPT database began to collect information. In addition, the database initially tracked only international terrorist incidents; it did not begin tracking domestic incidents until 1998. See National Memorial Institute for the Prevention of Terrorism, *MIPT Terrorism Knowledge Base: A Comprehensive Databank of Global Terrorist Incidents and Organizations*, Oklahoma City, Okla., ongoing since 2003.

The first step was to code the dependent variable: the end of terrorist groups. To compile the list of terrorist groups, we used the RAND-MIPT Terrorism Incident database, which has extensive coverage of terrorist groups beginning in the late 1960s. Consequently, our study started with 1968. Coding the ending of terrorist groups poses a methodological challenge, since it is never entirely clear when a terrorist group begins or ends. The start year of a terrorist group was assigned based on the first indication that the group existed and was dedicated to the use of violence. The end year of a terrorist group was assigned based on the earliest evidence that the group no longer existed or that the group no longer used terrorism to achieve its goals. This may be because security forces captured or killed most of its members, the group reached a peace agreement with the government, its members shifted to nonviolent means to achieve their goals, or its members splintered to join other groups or start new ones. Regardless of the reason, the group did not commit further terrorist attacks under its name.

The second step was to code each of the explanatory variables. We used the RAND-MIPT Terrorism Incident database as the starting point for many of the variables. For the breadth of goals, we provided a qualitative assessment of the group's goals, based on its statements and actions. For ideological motivation, we assessed whether groups were primarily religious, nationalist, right-wing, or left-wing. In such cases as Hamas, whose ideology is a combination of nationalism and religion, we assessed the primary motivation. We coded Hamas, for example, as primarily nationalist because its fundamental objective is the liberation of Palestine.[10] For regime type, we used data from Freedom House to code the regime type in states where terror groups had their base of support. We coded states as free, partly free, or not free.[11] For economic conditions, we used data from the World Bank, which

[10] On Hamas' goals, see Shaul Mishal and Avraham Sela, *The Palestinian Hamas: Vision, Violence, and Coexistence*, New York: Columbia University Press, 2000; and Khalid Harub, *Hamas: Political Thought and Practice*, Washington, D.C.: Institute for Palestine Studies, 2000.

[11] On the Freedom House codings, see Freedom House, *Freedom in the World 2007: The Annual Survey of Political Rights and Civil Liberties*, New York: Freedom House, 2007.

divided countries worldwide into four income groups according their per-capita income: low, lower middle, upper middle, and high.[12] For the size of the group, we used data from the RAND-MIPT Terrorism Incident database to divide groups into sizes based on the number of core members: fewer than 100; 100–999; 1,000–9,999; and 10,000 or more.

The third step was to assess the data. We took two approaches: (1) We assessed the primary reason for the group ending, and (2) we conducted a statistical analysis to assess the impact of economic conditions, regime type, size, ideology, and group goals.

The other methodological approach involved a comparative case-study approach to understand how specific terrorist groups ended.[13] Case studies offer a useful approach to help understand how and why groups ended.[14] What were the key factors? How did they cause the end of the group? This is virtually impossible to do without examining specific cases. As Alexander George and Timothy McKeown wrote, case studies are useful in uncovering

> what stimuli the actors attend to; the decision process that makes use of these stimuli to arrive at decisions; the actual behavior that then occurs; the effect of various institutional arrangements on

---

[12] On the World Bank codings, see World Bank, *World Development Indicators 2007*, Washington, D.C., 2007, p. xxi.

[13] In particular, see Alexander L. George, "Case Studies and Theory Development: The Method of Structured, Focused Comparison," in Paul Gordon Lauren, ed., *Diplomacy: New Approaches in History, Theory, and Policy*, New York: Free Press, 1979, pp. 43–68.

[14] On the costs and benefits of comparative case studies, see David Collier, "New Perspectives on the Comparative Method," in Dankwart A. Rustow and Kenneth Paul Erickson, eds., *Comparative Political Dynamics: Global Research Perspectives*, New York: HarperCollins, 1991, pp. 7–31; Charles C. Ragin, "Comparative Sociology and the Comparative Method," *International Journal of Comparative Sociology*, Vol. 22, Nos. 1–2, March–June 1981, pp. 102–120; Charles Tilly, "Means and Ends of Comparison in Macrosociology," in Fredrik Engelstad, ed., *Comparative Social Research*, Vol. 16: *Methodological Issues in Comparative Social Science*, London: JAI, 1997, pp. 43–53; Theda Skocpol and Margaret Somers, "The Uses of Comparative History in Macrosocial Inquiry," *Comparative Studies in Society and History*, Vol. 22, No. 2, April 1980, pp. 174–197; and Stephen Van Evera, *Guide to Methods for Students of Political Science*, Ithaca, N.Y.: Cornell University Press, 1997, pp. 49–88.

attention, processing, and behavior; and the effect of other variables of interest on attention, processing, and behavior.[15]

We then drew on the conclusions to inform our assessment of U.S. operations against al Qa'ida.

Outline of This Book

Chapter Two examines the universe of terrorist groups between 1968 and 2006 and asks, How have terrorist groups ended? Among those that have ended, did they end because a state's police or military forces defeated them? Did they end because they achieved victory? Or did they end for other reasons? This chapter looks at the 648 terrorist groups that existed between 1968 and 2006 and examines how groups ended.

Chapters Three, Four, and Five report a series of case studies. We used several criteria to select cases. First, we selected cases in which a terrorist group ended, so that we could examine the process in more detail. What policies were adopted? Were there other factors that contributed to the end of the group? Second, we wanted to examine some of the major reasons that groups end, especially politics, policing, and military force. Third, we wanted to analyze cases from different geographic regions. Consequently, Chapter Three examines the process of politics that led to the end of the Farabundo Marti National Liberation Front (Frente Farabundo Martí para la Liberación Nacional, or FMLN) in El Salvador. Chapter Four assesses the impact of policing on the end of Aum Shinrikyo in Japan, especially after the 1995 Tokyo subway attacks. And Chapter Five explores the impact of military force on al Qa'ida in Iraq.

Chapters Six and Seven turn to al Qa'ida. Chapter Six argues that al Qa'ida has been involved in more terrorist attacks in a wider

---

[15] Alexander L. George and Timothy J. McKeown, "Case Studies and Theories of Organizational Decision Making," in Robert F. Coulam and Richard A. Smith, *Advances in Information Processing in Organizations: A Research Annual*, Vol. 2, Greenwich, Conn.: JAI Press, 1985, pp. 21–58, p. 35.

geographical area since September 11, 2001, than it had during its previous history. These attacks spanned a geographic area across Europe, Asia, the Middle East, and Africa. The chapter then argues that the United States should fundamentally rethink its strategy toward al Qa'ida. Current efforts have not been successful, mostly because they have relied too much on military force. Chapter Seven argues that the United States should make police and intelligence efforts the backbone of U.S. counterterrorism policy and move away from its mantra of fighting a war on terrorism. This requires a paradigmatic shift.

CHAPTER TWO

How Terrorist Groups End

How do terrorist groups end? All terrorist groups eventually end. But do they end because police or military forces defeat them? Do they end by achieving victory? Or do they end for other reasons? As noted earlier, there has been little systematic work on how terrorist groups end. Understanding how terrorist groups have ended has significant implications for the U.S. struggle against such groups as al Qa'ida.

This chapter reviews the history of terrorist groups since 1968 and argues that terrorist groups usually end for two major reasons: They decide to adopt nonviolent tactics and join the political process, or local law-enforcement agencies arrest or kill key members of the group. Military force has rarely been the primary reason that terrorist groups end, and, since 1968, few groups have ended by achieving victory. This suggests that, where terrorist groups cannot or will not make a transition to nonviolence, policing is usually most effective in defeating terrorist groups. Police have a permanent presence in cities, towns, and villages; a better understanding of local communities than other security forces; and better intelligence. This means that they are best able to understand and penetrate terrorist networks. The use of military force has usually been effective in defeating terrorist groups when the terrorist group is so powerful that it becomes involved in an insurgency. We define an *insurgency* as an internal conflict in which (1) a group or groups are trying to overthrow the government or secede from it, (2) more than 1,000 have died over the course of the war, and (3) more than 100 have died on each side.

This chapter is divided into four sections. First, it summarizes the core arguments about how groups end. Second, it outlines the research design to test these arguments. Third, it examines the data and provides an overview of the findings. Fourth, it offers a brief conclusion.

The End of Terrorist Groups

This section begins by outlining the five main arguments about why terrorist groups end. It then outlines a range of other explanatory factors that may affect the end of terrorist groups.

How Groups End

There are at least five major ways in which terrorist groups end: policing, military force, splintering, politics, or victory.[1] Other tools may also be useful: providing economic aid to countries dealing with terrorism, imposing economic sanctions on states that harbor terrorist groups, dissuading groups by hardening targets, improving intelligence, or engaging in diplomacy.[2] But these are too weak to be used in a leading role. In practice, terrorist groups typically end due to a combination of factors. But with limited resources and attention, policymakers need to understand where to prioritize their efforts. Consequently, in coding how groups end, we assessed the primary ways in which they ended. Many factors may have contributed to the end of a group, but which was the most significant? What carried the most weight in pushing the group over the edge?

---

[1] We did not include a separate category for capturing or killing enemy leaders, what is often referred to as a *decapitation* strategy. Rather, we subsumed this under other categories. If a police force adopted this strategy, we included it under law enforcement. If military forces adopted it, we included it under military force. On economic aid, sanctions, and other counterterrorist instruments, see Audrey Kurth Cronin and James M. Ludes, eds., *Attacking Terrorism: Elements of a Grand Strategy*, Washington, D.C.: Georgetown University Press, 2004; and Paul R. Pillar, *Terrorism and U.S. Foreign Policy*, Washington, D.C.: Brookings Institution Press, 2001.

[2] Trager and Zagorcheva (2005–2006).

Some argue that groups break up primarily because police defeat them.[3] *Policing* involves the use of police and intelligence units to collect information on terrorist groups, penetrate cells, and arrest key members. A law-enforcement approach may also include developing antiterrorism legislation. This can involve criminalizing activities that are necessary for terrorist groups to function, such as raising money or recruiting openly. As one study concluded, "Terrorism is a crime, and therefore the primary mechanism of any liberal democracy in its efforts against it should be its criminal justice system."[4]

Others contend that terrorist groups can be defeated primarily by *military force*. As MIT political scientist Barry Posen argued, "Offensive action and offensive military capabilities are necessary components of a successful counterterror strategy."[5] This involves deploying military forces to capture or kill key members of the terrorist group. In some cases, it might include using military force against states that assist terrorist groups, to undermine their support base. It can involve the use of air strikes and standoff weapons against terrorists and those who harbor them, supported by special operations forces on the ground.[6] Even unsuccessful offensive actions that force terrorist units or cells to stay perpetually on the move to avoid destruction may help to reduce their capability. Constant surveillance makes it difficult for them to plan

---

[3] See, for example, Lindsay Clutterbuck, "Law Enforcement," in Audrey Kurth Cronin and James M. Ludes, eds., *Attacking Terrorism: Elements of a Grand Strategy*, Washington, D.C.: Georgetown University Press, 2004, pp. 140–161; Ross and Gurr (1989, pp. 409–409); Paul Wilkinson, *Terrorism and the Liberal State*, 2nd ed., Basingstoke, UK: Macmillan, 1986; Michael A. Sheehan, "Diplomacy," in Audrey Kurth Cronin and James M. Ludes, eds., *Attacking Terrorism: Elements of a Grand Strategy*, Washington, D.C.: Georgetown University Press, 2004, pp. 104–105.

[4] Clutterbuck (2004, p. 157).

[5] Barry Posen, "The Struggle Against Terrorism: Grand Strategy, Strategy, and Tactics," *International Security*, Vol. 26, No. 3, Winter 2001–2002, p. 39–55, p. 47.

[6] Timothy D. Hoyt, "Military Force," in Audrey Kurth Cronin and James M. Ludes, eds., *Attacking Terrorism: Elements of a Grand Strategy*, Washington, D.C.: Georgetown University Press, 2004, pp. 162–185; Cronin (2002–2003, p. 55); Trager and Zagorcheva (2005–2006, pp. 118–120); David C. Rapoport, "The Four Waves of Modern Terrorism," in Audrey Kurth Cronin and James M. Ludes, eds., *Attacking Terrorism: Elements of a Grand Strategy*, Washington, D.C.: Georgetown University Press, 2004, pp. 46–73, p. 60.

and organize. Constant pursuit makes it dangerous for them to rest. In short, this argument assumes that even the threat of military force may be critical to exhaust terrorists and deter their state sponsors.[7] Military force can include intervening abroad, such as the U.S. operations in Afghanistan in 2001 or Israeli operations in Lebanon in 1982.[8] It can also include intervening domestically, such as Turkish military incursions against the Kurdistan Workers' Party (Partiya Karkerên Kurdistan, or PKK) or Russian military actions against Chechen groups.[9] States generally use military force domestically when a terrorist group is involved in an insurgency and seeks to overthrow the government or secede from it.[10]

As one study concluded: "The use of military force has hastened the decline or ended a number of terrorist groups, including the late–19th-century Russian group Narodnaya Volya, Peru's Shining Path, and Kurdistan Workers' Party."[11] Military force was critical to the U.S. response after the September 11, 2001, attacks. As President George W. Bush stated in response to the attacks, the primary counterterrorist response was a military one:

> While the most visible military action is in Afghanistan, America is acting elsewhere. We now have troops in the Philippines,

---

[7] Posen (2001–2002).

[8] David Rapoport argued that military defeat of terrorists in such countries as Lebanon caused the end of what he called the *third wave* of terrorism in the 1980s. See David C. Rapoport, "The Fourth Wave: September 11 in the History of Terrorism," *Current History*, Vol. 100, December 2001, pp. 419–424, p. 421; David C. Rapoport, "Terrorism," in Lester R. Kurtz and Jennifer E. Turpin, eds., *Encyclopedia of Violence, Peace, and Conflict*, San Diego, Calif.: Academic Press, 1999, pp. 497–510; and Rapoport (2004, pp. 3–4).

[9] Cronin (2006, p. 30); Mark Kramer, "The Perils of Counterinsurgency: Russia's War in Chechnya," *International Security*, Vol. 29, No. 3, Winter 2004–2005, pp. 5–62.

[10] An insurgency is a political-military campaign by nonstate actors seeking to overthrow a government or secede from a country through the use of unconventional—and, sometimes, conventional—military strategies and tactics. On the definition of insurgency, see *Guide to the Analysis of Insurgency* (1986, p. 2) and U.S. Joint Chiefs of Staff, *Department of Defense Dictionary of Military and Associated Terms*, Washington, D.C., joint publication 1-02, ongoing since 1972, p. 266.

[11] Cronin (2006, p. 30).

> helping to train that country's armed forces to go after terrorist cells that have executed an American, and still hold hostages. Our soldiers, working with the Bosnian government, seized terrorists who were plotting to bomb our embassy. Our Navy is patrolling the coast of Africa to block the shipment of weapons and the establishment of terrorist camps in Somalia.[12]

Groups may also break up as a result of competition among terrorist groups, what can be called splintering. Terrorists sometimes calculate that they have a better chance of reaching their objectives if they join a stronger group or start a new one. Most terrorist groups conduct some form of implicit cost-benefit analysis. They need a support base that provides needed material, such as money, safe houses, and recruits. They also require a hospitable environment to survive. Terrorist groups often compete with each other for these resources and support.[13] In Palestinian territory, for example, there is a range of groups, such as Palestinian Islamic Jihad, Hamas, Popular Front for the Liberation of Palestine–General Command, and a variety of Fatah organizations, such as al-Aqsa Martyrs' Brigades. In Pakistan, a plethora of groups, such as Jaish-e-Mohammad (JeM) and Lashkar-e-Taiba, compete for resources and support. Groups that fail to secure a sufficient amount of support may break apart as members scatter to bigger and more powerful groups.

The critical issue for splintering is that the end of a group does not signal the end of terrorism by its members. Members remain committed to terrorism but choose to continue fighting for other groups. Consequently, in the next section, which assesses how terrorist groups end, we focus on cases in which (1) the terrorist group ends and (2) most of its members stop using terrorism. We thus exclude splintering.

In addition, groups may end the use of terrorism because its members view nonviolent political means as more effective to achieve its goals, what can be called politics. Nonviolent alternatives to terrorism usually involve cooperation with the government on a collective

---

[12] George W. Bush, *State of the Union*, Washington, D.C.: White House, 2002.

[13] Bloom (2005, pp. 76–100); Kydd and Walter (2006, pp. 58, 76–78).

or individual level.[14] In some cases, groups may choose to participate in politics following a peace settlement with the government. This could be because structural conditions have changed, such as a transition from authoritarianism to democracy. In addition, nonviolent groups may temporarily use terrorism in response to a specific event or events and then return to nonviolence. In France, for example, left-wing groups, such as the Totally Anti-War Group (Association Totalement Anti-Guerre, or ATAG), briefly resorted to terrorism to protest the U.S. war in Afghanistan in 2001 but then returned to nonviolence. Some argue that a wide range of variables can affect a group's willingness to turn to nonviolent means, such as the group's organizational structure (hierarchical organizations may be more likely to negotiate than networked groups) and the nature of public support for the cause (groups with more-ambivalent support networks may be more likely to compromise).[15]

Finally, some groups end once they attain *victory*.[16] That is, terrorist groups may abandon terrorism because their objectives have been achieved. There are several historical examples: the Irgun Zvai Leumi in Israel, the Ethniki Organosis Kyprion Agoniston (EOKA) in Cyprus, and National Liberation Front (Front de Libération Nationale, or FLN) in Algeria. As Bruce Hoffman argued,

> Although governments throughout history and all over the world always claim that terrorism is ineffective as an instrument of political change, the examples of Israel, Cyprus, and Algeria . . . provide convincing evidence to the contrary.[17]

Some terrorist groups, such as the Irgun, EOKA, and FLN, may play a direct role in achieving victory, though terrorism alone is rarely suf-

---

[14] Crenshaw (1996, pp. 266–268); USIP (1999).

[15] Cronin (2006, pp. 25–27).

[16] Ross and Gurr (1989, p. 408); Dershowitz (2002); Pape (2005); Lake (2002); Kydd and Walter (2006); USIP (1999).

[17] Hoffman (2006, p. 61).

ficient to achieve success.[18] In other cases, a group may play little or no role in causing change, but the change they seek happens anyway. For instance, a number of terrorist groups that advocated the creation of an independent Armenian state, such as the Armenian Resistance Group, disbanded after the collapse of the Soviet Union. While they played no role in the collapse of the Soviet Union, one of their major objectives (independence for Armenia) was, in fact, achieved.

Explanatory Variables

We also looked at five factors that can influence how long terrorist groups last and how they end: ideological motivation, economic conditions, regime type, the size of groups, and the breadth of terrorist goals. Governments cannot control most of these factors.

A group's *ideological motivation* may influence how the group ends and after how many years. Groups can be divided into at least four types: left-wing, right-wing, nationalist, or religious.[19] Left-wing includes a range of Marxist-Leninist, environmental, animal rights, anarchical, and antiglobalization groups. Right-wing includes racist and fascist groups. Nationalist includes groups inspired by a desire for independence, territorial control, or autonomy because of ethnic or other affiliations. Religious terrorists commit acts of terrorism to comply with a religious mandate or to force others to follow that mandate.

Some studies have suggested that terrorist groups motivated by nationalist and religious goals last the longest.[20] They typically have strong sources of support among the local population of the same ethnicity, and "broader popular support is usually the key to the greater average longevity of ethno-nationalist/separatist groups in the modern era."[21] As another study concluded,

> The nature of the grievances matters. Ethnically based terrorist campaigns can be harder to end decisively than politically based

---

[18] Crenshaw (1996, p. 260).

[19] Cronin (2002–2003, pp. 39–42).

[20] Cronin (2002–2003; 2006, p. 13); USIP (1999).

[21] Cronin (2002–2003, p. 40).

> ones, because they often enjoy broader support among a population they seek to represent.[22]

Some have also argued that religious groups have long durations because of the staying power of sacred or spiritually based motivations.[23] Similar research suggests that left-wing and right-wing groups have the shortest durations. The logic is that they frequently have trouble identifying concrete goals and retaining popular support. As Audrey Kurth Cronin noted,

> The left-wing groups of the 1970s, for example, were notorious for their inability to articulate a clear vision of their goals that could be handed down to successors after the first generation of radical leaders departed or were eliminated.[24]

Martha Crenshaw argued that right-wing groups have trouble retaining popular support, partly because they often have such decentralized organizational structures that generational transition becomes extremely challenging.[25]

Economic conditions may also affect terrorist groups. Poor economic conditions may heighten grievances, which provide a more supportive environment for terrorist groups and increase their longevity. Grievances are difficult to measure independently of terrorism, but measures of average levels of discrimination are feasible. Some argue that greater economic inequality creates broad grievances that favor terrorism.[26] "Governments that fail to meet the basic welfare and eco-

---

[22] USIP (1999, p. 1).

[23] David C. Rapoport, "Fear and Trembling: Terrorism in Three Religious Traditions," *American Political Science Review*, Vol. 78, No. 3, September 1984, pp. 658–677; Cronin (2006, p. 13). On the link between economic conditions and terrorism, also see Michael Mousseau, "Market Civilization and Its Clash with Terror," *International Security*, Vol. 27, No. 3, Winter 2002–2003, pp. 5–29.

[24] Cronin (2006, p. 23).

[25] USIP (1999, p. 78); Cronin (2002–2003, pp. 39–42).

[26] On civil conflict, see Edward N. Muller, "Income Inequality, Regime Repressiveness, and Political Violence," *American Sociological Review*, Vol. 50, No. 1, February 1985, pp. 47–61.

nomic needs of their peoples and suppress their liberties," wrote Samuel Huntington, "generate violent opposition to themselves and to Western governments that support them."[27] This means that terrorist groups tend to last longer in poor countries. If so, one way to end terrorism is to improve the economic condition of countries where terrorism is most prevalent.[28]

Regime type may also be correlated with the duration of terrorist groups. For instance, some argue that democracy is associated with less discrimination and repression along cultural or other lines, since it endows citizens with a political power (the vote) that they do not enjoy in dictatorships. Even more directly, how well states observe civil rights—such as freedom of association, expression, and due process—should be associated with lower grievances.[29] Others have argued the reverse: Democratic systems place significant constraints on state behavior. As one study concluded,

> Democracies may be more constrained in their ability to retaliate than authoritarian regimes. . . . Capable authoritarian regimes are able to gather more information on their populations than democracies and can more easily round up terrorists and target those sympathetic to them.[30]

The size of terrorist groups may also affect duration. Larger groups may be able to last longer than smaller groups. Other things being equal, they should have more resources, which allows them to sustain activities longer than smaller groups. In addition, it may take more time for governments to break up larger groups, since they have more mem-

---

[27] Samuel P. Huntington, "The Age of Muslim Wars," *Newsweek*, December 17, 2001, p. 48.

[28] Mousseau (2002–2003, pp. 24–26).

[29] Lower grievances, however, do not necessarily translate into less likelihood of war or terrorism. See, for example, James D. Fearon and David D. Laitin, "Ethnicity, Insurgency, and Civil War," *American Political Science Review*, Vol. 97, No. 1, February 2003, pp. 75–90, pp. 84–85.

[30] Kydd and Walter (2006, p. 61); also see Pape (2005, p. 44).

bers and resources. This may be especially true for insurgent groups, which frequently have greater resources to fight the government.

Finally, some have argued that the *breadth* of terrorist goals may be linked to the end of terrorist groups.[31] Terrorist goals can range from narrow ones (such as coercing a government to change a specific policy) to broader ones (such as overthrowing multiple regimes). First, groups may seek the status quo and support an existing regime or territorial arrangement and oppose groups that seek to change it. Examples include the United Self-Defense Forces of Colombia and loyalist paramilitary groups in Northern Ireland, such as the Ulster Volunteer Force. Second, groups may seek policy change and advocate a range of policy demands. Examples include a desire to remove U.S. troops from the Middle East or protest U.S. policies there or improve the treatment of animals.[32] Third, groups may seek territorial change and wish to acquire territory from a state to establish a new state, gain autonomy, or join another state. Fourth, groups may seek regime change and wish to overthrow a government and replace it with one led by the terrorists—or at least with one more to their liking. This can also include independence from a colonial power. Fifth, groups may seek what might be called *empire*. They seek to overthrow more than one regime and establish a single, sovereign authority, such as a caliphate. Sixth, groups may seek social revolution and wish to alter individuals' social or cultural norms. Examples include the agendas of some racist groups or the aims of some Marxist-Leninist groups.

Summary of Results

The results of the data analysis were stark. Terrorist groups end for two major reasons: Members decide to adopt nonviolent tactics and join the political process (43 percent), or local law-enforcement agencies arrest

---

[31] See, for example, Max Abrahms, "Why Terrorism Does Not Work," *International Security*, Vol. 31, No. 2, Fall 2006, pp. 42–78; and Kydd and Walter (2006).

[32] Groups were also coded as seeking policy change if they committed terrorist acts to protest a policy, such as U.S. foreign policy.

or kill key members of the group (40 percent). Military force has rarely been the primary reason for the end of terrorist groups (7 percent), and few groups since 1968 have achieved victory (10 percent).[33] Figure 2.1 illustrates the results. A large number of groups also end by splintering, but their members continue to pursue terrorism by joining an existing group or creating a new one. Since terrorism continues, we focus only on those cases in which a terrorist group ends and most of its members

Figure 2.1
How Terrorist Groups End (N = 268)

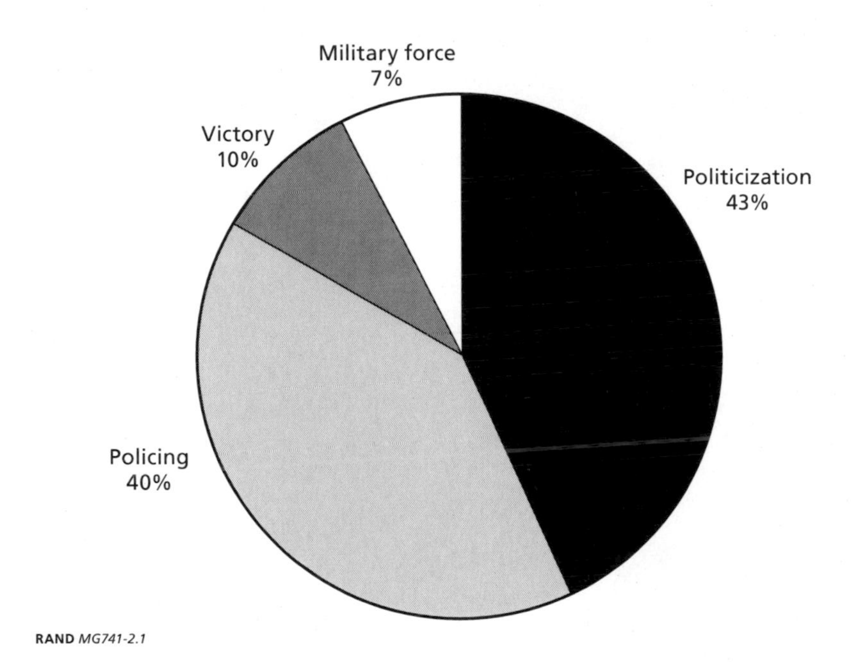

[33] In generating these percentages, we excluded from the original 648 groups (1) those that were still active (244 groups) and (2) those that ended because of splintering, since the members still used terrorism (136 groups). Out of the 268 remaining groups, 20 ended by military force (7 percent), 27 by victory (10 percent), 107 by policing (40 percent), and 114 by politics (43 percent).

stop using terrorism as a tactic.[34] The next two sections explore the logic of politics and policing in more detail.

Political Solutions

Most terrorist groups that have ended did so by pursuing their goals through politics. The possibility of a political solution is linked to a key variable: the breadth of terrorist goals. As Figure 2.2 shows, most terrorist groups that end because of politics seek narrow policy goals, such as policy change, territorial change, or regime change. The further right on the x-axis, the less likely a terrorist group will end because of politics. There are two major logics at play. First, the narrower a

Figure 2.2
Politics and Group Goals

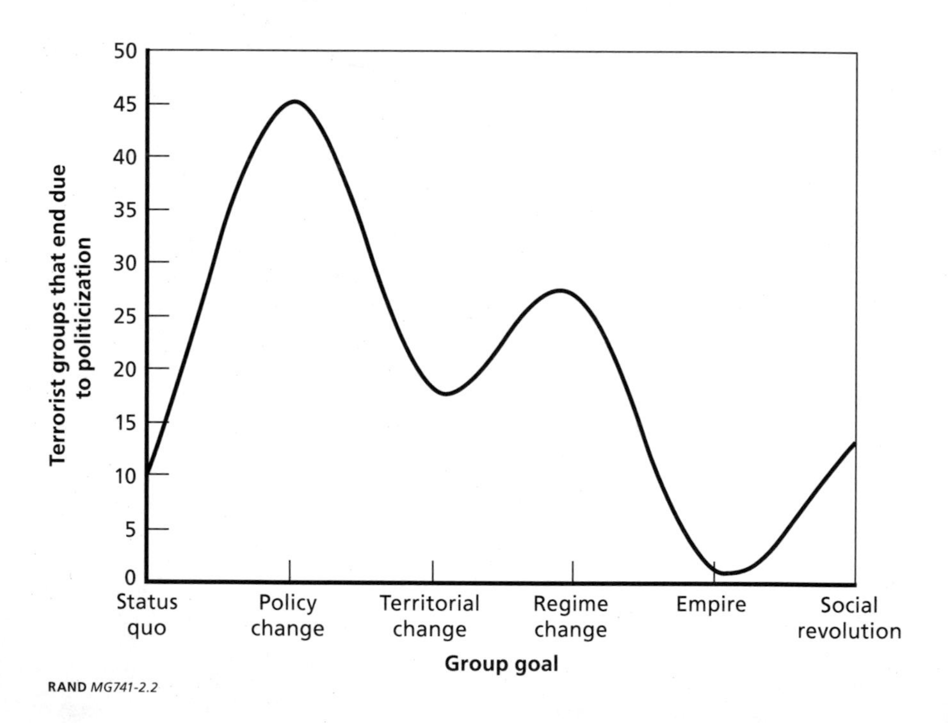

RAND *MG741-2.2*

---

[34] Out of the 648 total groups active since 1968, 244 terrorist groups still existed in 2006, and 404 had ended. Of the 404 groups that ended, 136 (or 28 percent) ended because of splintering.

terrorist organization's goals, the more likely the government and terrorist group are to be able to agree on a settlement. Second, the narrower the goals, the more difficult it is for terrorists to achieve them, and the more willing they will be to seek nonviolent means.

One path to politics includes an explicit *peace settlement.* It involves a bargaining process with the government in which terrorist groups reach an agreement to disband their militant wings and transition to a political party.[35] This could be for a number of reasons. One is because structural conditions have changed, such as the loss of outside state support. The collapse of the Soviet Union meant that a number of groups, such as the FMLN in El Salvador, saw their outside assistance quickly begin to dry up. Or it could be because the two sides have reached a stalemate. "If," Paul Pillar wrote, "the stakes are indivisible, so that neither side can get most of what it wants without depriving the other of most of what it wants, negotiations are less apt to be successful."[36] Terrorism may persist in a country precisely because the goals that the government and the terrorist group are pursuing are so far apart, "with nothing in between to contribute to the give and take of negotiation and bargaining."[37] Terrorists fighting for broad goals, such as social revolution or empire, are less likely to reach a negotiated settlement than are groups fighting for limited aims, such as policy change or territorial change. Where a terrorist group's goals are minimal, there may be a middle ground from which to draw a compromise settlement.

As Figure 2.3 highlights, terrorist goals can range from narrow ones (such as securing a policy change) to broader ones (such as changing a country's social order). The further right on the x-axis, the broader the goals, beginning with the status quo and moving through policy

---

[35] Crenshaw (1996, pp. 266–268); USIP (1999).

[36] Paul R. Pillar, *Negotiating Peace: War Termination as a Bargaining Process*, Princeton, N.J.: Princeton University Press, 1983, p. 24.

[37] I. William Zartman, "The Unfinished Agenda: Negotiating Internal Conflicts," in Roy E. Licklider, ed., *Stopping the Killing: How Civil Wars End*, New York: New York University Press, 1993, pp. 25–26; also see Stephen John Stedman, "Negotiation and Mediation in Internal Conflict," in Michael E. Brown, ed., *The International Dimensions of Internal Conflict*, Cambridge, Mass.: MIT Press, 1996, pp. 341–376.

Figure 2.3
Settlement Between States and Terrorist Groups

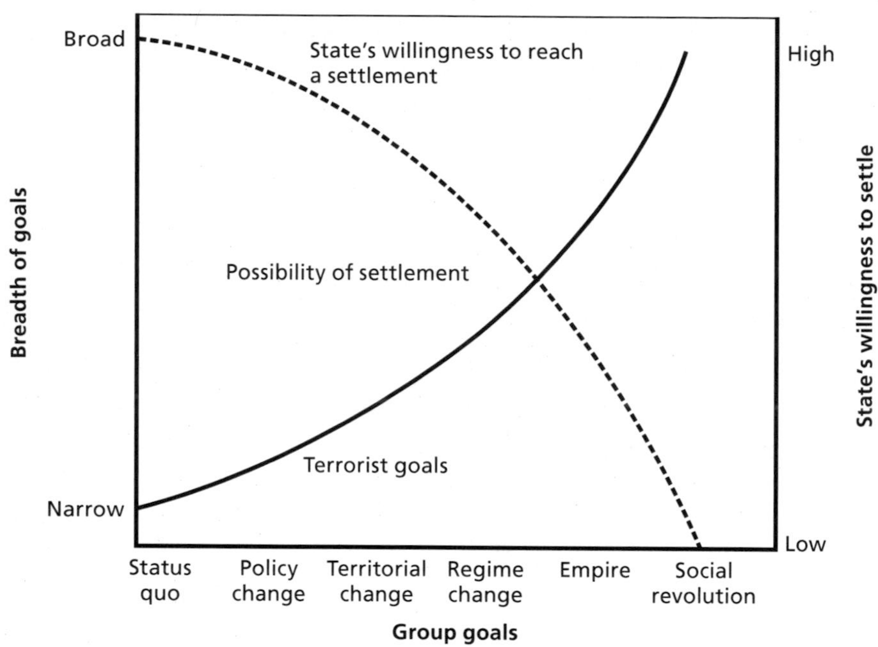

RAND *MG741-2.3*

change, territorial change, regime change, empire, and broad social revolution. At the same time, we should also expect that a state's willingness to bargain with a terrorist group declines as the group's goals become broader. The reason is simple: The government has more to lose. Agreeing to change policy is easier than agreeing to overturn an entire social order. *The Turner Diaries*, a bedrock text for right-wing groups in the United States, called for social revolution. It began, "September 16, 1991. Today it finally began! After all these years of talking—and nothing but talking—we have finally taken our first action. We are at war with the System, and it is no longer a war of words."[38]

There are several examples of political settlement occurring between a government and terrorist group. One is the IRA, which

---

[38] Andrew Macdonald, *The Turner Diaries*, 2nd ed., New York: Barricade Books, 1996. William Luther Pierce wrote the book under this pseudonym.

ended its terrorist activity following negotiations with the United Kingdom and Republic of Ireland. The Belfast (Good Friday) Agreement,[39] which was announced on April 10, 1998, addressed the main issues of internal governance and detailed measures concerning constitutional changes, decommissioning, security, and paramilitary prisoners.[40] In El Salvador, the FMLN began to disarm in the aftermath of the January 1992 Chapultepec Peace Accords.[41] The agreement outlined reforming the armed forces and police into a legal political party.[42] In Mozambique, the Resistencia Nacional Mozambicana (RENAMO) signed a peace agreement with the government in October 1992, which included a cease-fire, disarmament and demobilization process, and multiparty elections.[43] It won 112 seats in the national assembly in the October 1994 elections. Finally, the April 19 Movement (Movimiento 19 de Abril, or M-19) in Colombia negotiated a settlement with the Colombian government in 1989 and participated in discussions to draw up a new constitution. The government offered M-19 participa-

---

[39] Great Britain Northern Ireland Office, *The Belfast Agreement: An Agreement Reached at the Multi-Party Talks on Northern Ireland*, London: Stationery Office, 1998.

[40] See, for example, Douglas Woodwell, "The 'Troubles of Northern Ireland': Civil Conflict in an Economically Well-Developed State," in Paul Collier and Nicholas Sambanis, eds., *Understanding Civil War: Evidence and Analysis*, Vol. 2: *Europe, Central Asia, and Other Regions*, Washington, D.C.: World Bank, 2005, pp. 161–190.

[41] Embassy of El Salvador, "The Peace Accords," undated Web page.

[42] Michael W. Doyle, Ian Johnstone, and Robert C. Orr, eds., *Keeping the Peace: Multidimensional UN Operations in Cambodia and El Salvador*, New York: Cambridge University Press, 1997; Elisabeth Jean Wood, *Forging Democracy from Below: Insurgent Transitions in South Africa and El Salvador*, Cambridge and New York: Cambridge University Press, 2000; Charles T. Call, "Assessing El Salvador's Transition from Civil War to Peace," in Stephen John Stedman, Donald S. Rothchild, and Elizabeth M. Cousens, *Ending Civil Wars: The Implementation of Peace Agreements*, Boulder, Colo.: Lynne Rienner, 2002, pp. 383–420; Tommie Sue Montgomery, *Revolution in El Salvador: From Civil Strife to Civil Peace*, 2nd ed., Boulder, Colo.: Westview Press, 1995.

[43] Chris Alden, *Mozambique and the Construction of the New African State: From Negotiations to Nation Building*, New York: Palgrave, 2001; United Nations Department of Public Information, *The United Nations and Mozambique, 1992–1995*, New York, 1995b.

tion in the political system and a role in forming a political party.[44] In each of these cases, the group had narrow-enough goals that a negotiated settlement was possible.

The second logic can best be described as a transition to *civic action*. Unlike the peace-settlement option, terrorist groups that pursue this path do not necessarily reach a formal agreement with the government. In some cases, the government may grant amnesty to terrorists. In Italy, for example, the government established a policy of repentance in the 1980s, which provided "the extension of leniency in return for disassociation."[45] The Italian policy encouraged individual exit from terrorist groups. The Italian government offered reduced prison sentences in exchange for information that would enable the government to dismantle terrorist groups. By 1989, 389 terrorists had repented.[46]

In general, we should expect that, the narrower the goals, the more difficult it is for terrorists to achieve them, and the more likely terrorist groups may be willing to seek nonviolent means. This is especially true where terrorist groups are small and their goals sufficiently narrow. In such cases, pursuing their goals through nonviolence has greater benefits and lower costs. Indeed, most terrorist organizations are implicitly strategic calculators.[47] They use violence to achieve specific political purposes, such as coercing a target government to change policy, mobilizing additional recruits and financial support, or achieving independence. They usually have a set of hierarchically ordered goals and choose strategies that best advance them.

---

[44] María Eugenia Vásquez Perdomo, *My Life as a Colombian Revolutionary: Reflections of a Former Guerrillera*, Philadelphia, Pa.: Temple University Press, 2005; Americas Watch Committee, *The Killings in Colombia*, Washington, D.C., 1989.

[45] Leonard Weinberg, and William Lee Eubank, *The Rise and Fall of Italian Terrorism*, Boulder, Colo.: Westview Press, 1987, p. 129.

[46] Weinberg and Eubank (1987).

[47] See Crenshaw (1990, p. 117); Pape (2003); Jonathan Schachter, *The Eye of the Believer: Psychological Influences on Counter-Terrorism Policy-Making*, Santa Monica, Calif.: RAND Corporation, RGSD-166, 2002, p. 96; and Andrew Kydd and Barbara F. Walter, "Sabotaging the Peace: The Politics of Extremist Violence," *International Organization*, Vol. 56, No. 2, Spring 2002, pp. 263–296, pp. 279–289.

Consequently, they are influenced by cost-benefit calculations. Resorting to terrorism has benefits if a group can successfully achieve its goals. But it also has costs. Terrorists must be constantly underground and on the run, because government security forces are trying to capture or kill them. Terrorism provokes repression that some organizations believe they cannot survive. As Martha Crenshaw noted, "some revolutionaries perceive government strength as an obstacle to using terrorism."[48] Hans-Joachim Klein, a former West German terrorist, wrote that the experience of participating in a terrorist action (seizing ministers from the Organization of the Petroleum Exporting Countries [OPEC] in Vienna in 1975) convinced him to abandon the underground. He was tired of being constantly on the run and living covertly, and he could not adjust to a set of beliefs he regarded as callous and cynical with regard to human life.[49]

This is why politics cannot always be viewed as separate from policing. In some cases, opportunities for collective action, such as mass protest, may occur independently of government actions. But government action may also be a catalyst. An effective coercive policy against terrorist groups can make terrorism too dangerous and unproductive to continue—even in cases in which government security forces are unable to capture or kill key members of the group. The government response can contribute to the internal costs of terrorism by provoking organizational disagreements. Other costs associated with terrorism may be more significant than the direct or indirect penalties that governments can exact, especially in democracies. One of the most important costs is the withdrawal of popular support. The attitude of an initially sympathetic community on which any terrorist organization depends may change as a result of terrorist actions.[50] This is sometimes referred to as *backlash*.[51]

---

[48] Crenshaw (1996, p. 254).

[49] Hans-Joachim Klein, *La Mort Mercenaire: Témoignage d'un Ancient Terroriste Oust-Allemand*, Paris: Seuil, 1980.

[50] Crenshaw (1996, pp. 262–266).

[51] Ross and Gurr (1989); Ted Robert Gurr, "Terrorism in Democracies: Its Social and Political Bases," in Walter Reich, ed., *Origins of Terrorism: Psychologies, Ideologies, Theologies, States*

We should thus expect that, the broader the goals of terrorist groups, the less likely they are able to achieve them, and the more willing they are to use nonviolence to achieve them. There are numerous examples of groups that transitioned from terrorism to civic action because of changes in their assessment of costs and benefits. For example, a French group that called itself Gracchus Babeuf (after the 18th-century French philosopher) committed several terrorist attacks in France in 1990 and 1991 to protest U.S. actions in the first Persian Gulf War, as well as to protest U.S. and French policy toward Libya. In 1995, the Anti-Imperialist Group Liberty for Mumia Abu Jamal bombed a Chrysler dealership in Germany to protest the arrest of the African-American journalist Mumia Abu-Jamal, who was imprisoned for allegedly murdering a police officer.[52] In the 1980s, an environmental extremist group called the Peace Conquerors conducted several terrorist attacks in western Europe and Australia to coerce several countries and companies to change their environmental policies.[53] In France, left-wing groups, such as ATAG, briefly resorted to terrorism to protest the U.S. war in Afghanistan in 2001 but then returned to nonviolence. In all of these cases, group members dropped the use of terrorist tactics and switched to nonviolent civic action. Anecdotal evidence from historical cases supports this argument. For example, the 19th-century anarchist movement turned from violence to nonviolence (including the advocating of general strikes) as the working classes became more active and as governments and society become more tolerant of worker protests.[54]

---

of Mind, Washington, D.C.: Woodrow Wilson International Center for Scholars, 1990, pp. 86–102.

[52] See, for example, U.S. Department of State, Office of the Secretary of State, Office of the Coordinator for Counterterrorism, *Patterns of Global Terrorism, 1995*, Washington, D.C., 1996.

[53] Margaret Kosal, "Terrorism Targeting Industrial Chemical Facilities: Strategic Motivations and the Implications for U.S. Security," *Studies in Conflict and Terrorism*, Vol. 29, No. 7, October 2006, pp. 719–751.

[54] Crenshaw (1996, p. 266).

Policing

For terrorist groups that cannot or will not abandon terrorism, policing is likely to be the most effective strategy to destroy terrorist groups. The logic is straightforward: Police generally have better training and intelligence to penetrate and disrupt terrorist organizations. They are the primary arm of the government focused on internal security matters.[55] The mission of the police and other security forces should be to eliminate the terrorist organization—the command structure, terrorists, logistical support, and financial and political support—from the midst of the population. As Bruce Hoffman argued,

> Law enforcement officers should actively encourage and cultivate cooperation by building strong ties with community leaders, including elected officials, civil servants, clerics, businessmen, and teachers, among others, and thereby enlist their assistance and support.[56]

A police approach may also include developing antiterrorism legislation. This can involve criminalizing activities that are necessary for terrorist groups to function, such as raising money or openly recruiting. This means treating terrorism as a crime. Therefore, a state's primary mechanism for dealing with a crime is through its criminal-justice system.[57] The policing approach can also include providing foreign assistance to police and intelligence services abroad to improve their counterterrorism capacity.

During the Cold War, the United States provided assistance to foreign police and intelligence services to prevent countries from falling under Soviet influence.[58] By the early 1970s, the U.S. Congress became

---

[55] Roger Trinquier, *Modern Warfare: A French View of Counterinsurgency*, New York: Praeger, 1964, p. 43; David Galula, *Counterinsurgency Warfare: Theory and Practice*, St. Petersburg, Fla.: Hailer Publishing, 2005, p. 31.

[56] Hoffman (2006, p. 169).

[57] Clutterbuck (2004, p. 157).

[58] John Lewis Gaddis, *Strategies of Containment: A Critical Appraisal of Postwar American National Security Policy*, New York: Oxford University Press, 1982; H. W. Brands, *The Devil*

deeply concerned that U.S. assistance abroad frequently strengthened the recipient government's capacity for repression.[59] Congress was also concerned about the role of the Central Intelligence Agency (CIA), which trained foreign police in countersubversion, counterguerrilla, and intelligence-gathering techniques.[60] Consequently, Congress adopted §660 of the Foreign Assistance Act in 1974.[61] It prohibited the United States from providing internal security assistance to foreign governments and specifically stated that the U.S. government could not

> provide training or advice, or provide any financial support, for police, prisons, or other law enforcement forces for any foreign government or any program of internal intelligence or surveillance on behalf of any foreign government within the United States or abroad.[62]

Although the U.S. government still provided some internal security training during the late 1970s and 1980s through exemptions and waiver provisions, §660 largely halted U.S. involvement in this area. One notable exception was the International Criminal Investigative Training Assistance Program (ICITAP) in the U.S. Depart-

---

We Knew: Americans and the Cold War, New York: Oxford University Press, 1993.

[59] Michael McClintock, *The American Connection*, London: Zed Books, 1985; Martha Knisely Huggins, *Political Policing: The United States and Latin America*, Durham, N.C.: Duke University Press, 1998.

[60] John F. Kennedy, "Special Message to the Congress on the Defense Budget," March 28, 1961, in John F. Kennedy, *Public Papers of the Presidents of the United States: John F. Kennedy; Containing the Public Messages, Speeches, and Statements of the President, January 20 to December 31, 1961*, Washington, D.C.: U.S. Government Printing Office, 1962a, p. 236.

[61] Public Law 93-669, Foreign Assistance Act Amendments of 1974.

[62] U.S. Senate Committee on Foreign Relations and U.S. House of Representatives Committee on International Relations, *Legislation on Foreign Relations Through 2000*, Washington, D.C.: U.S. Government Printing Office, 2001, pp. 338–339; Robert Perito, *The American Experience with Police in Peace Operations*, Clementsport, N.S.: Canadian Peacekeeping Press, 2002, pp. 18–19.

ment of Justice, which was established in 1986 to help restructure the law-enforcement system of countries in transition.[63]

The end of the Cold War and the increasing tempo of U.S. stability operations after 1989 rendered the 1974 legislation largely obsolete. Section 660 still exists. But U.S. government agencies secured waivers and provided police and other internal security assistance to a range of democratic and nondemocratic regimes. U.S. assistance to foreign police for counterterrorism purposes was limited but did include several types: providing arms and other equipment; training and mentoring security forces; and building infrastructure, such as prisons and police stations. This assistance was geared toward promoting U.S. security and interests abroad by improving foreign governments' ability to deal with common security threats, especially terrorism. As the Foreign Assistance Act noted, counterterrorism assistance is critical "to enhance the ability of . . . law enforcement personnel to deter terrorists and terrorist groups from engaging in international terrorist acts such as bombing, kidnapping, assassination, hostage taking, and hijacking."[64] U.S. policymakers believed that strengthening the capabilities of foreign governments has a feedback loop: Improving foreign governments' ability to deal with security threats, such as terrorism, increases U.S. security.

There are numerous examples of policing successes against terrorist groups. Stepped-up surveillance by the Royal Canadian Mounted Police and the Quebec Provincial Police (QPP) led to a series of arrests and convictions of Quebec terrorist groups in the 1970s, especially the Liberation Front of Quebec (Front de Libération du Québec, or FLQ). As one study concluded,

> Stepped-up surveillance by the Royal Canadian Mounted Police Security Service (RCMPSS) and the Quebec Provincial Police (QPP), including the use of informants, led to arrests and convic-

---

[63] Charles T. Call, "Institutional Learning Within ICITAP," in Robert B. Oakley, Michael J. Dziedzic, and Eliot M. Goldberg, eds., Policing the New World Disorder: Peace Operations and Public Security, Washington, D.C.: National Defense University Press, 1998, pp. 315–364.

[64] P.L. 93-669, Chapter 8, Part 2.

> tions from late 1970 to 1972. These, along with the incarceration of previous members of the FLQ and the flight of other activists into foreign exile, coincided with the end of violent activism in 1972.[65]

So did increased Federal Bureau of Investigation (FBI) and state and local police activity against Puerto Rican, black-liberation, and white-supremacist groups in the 1970s and 1980s. As one study concluded, there is unequivocal evidence of FBI and local police "success in infiltrating militant organizations, preempting attacks, and arresting terrorist of every political stripe."[66] The Black Liberation Army, established in 1971 by embittered former members of the Black Panther Party, was responsible for about 20 ambushes of police officers. By 1974, 18 of its members were in prison, and only a handful of subsequent events were attributed to the survivors. In the early 1980s, the United Freedom Front claimed responsibility for roughly ten bombings against corporate and military targets in the New York City metropolitan area. All seven of its known members were arrested in 1984 and 1985.[67]

Other Reasons

There are several other reasons that terrorist groups might end, but they tend to be less common.

Military Force

Seven percent of terrorist groups that have ended since 1968 have done so because of military force. When they became strong enough to conduct insurgencies, however, terrorist groups ended because of military force 25 percent of the time.

---

[65] Ross and Gurr (1989, p. 412).

[66] Ross and Gurr (1989, p. 417).

[67] Ross and Gurr (1989, p. 41); see also Christopher Hewitt, *Understanding Terrorism in America: From the Klan to al Qaeda*, New York: Routledge, 2002.

Under most conditions, there are limits to the use of military force against terrorist groups. Most groups are small. Our data showed that nearly two-thirds of all terrorist groups active since 1968 have fewer than 100 members, making it difficult to engage them with large, conventional forces. Military forces may be able to penetrate and garrison an area that terrorist groups frequent and, if well sustained, may temporarily reduce terrorist activity. But once the situation in an area becomes untenable for terrorists, they will simply transfer their activity to another area, and the problem remains unresolved.[68] Terrorists groups generally fight wars of the weak. Most do not put large, organized forces into the field unless they have become involved in an insurgency.[69]

In addition, military force is usually too blunt an instrument for countering terrorism. Military tools have increased in precision and lethality, especially with the growing use of precision standoff weapons and imagery to monitor terrorist movement. These capabilities may limit the footprint of U.S. or other forces and minimize the costs and risks of sending in military forces to potentially hostile countries.[70] But even precision weapons have been of marginal use against terrorist groups. For example, the United States launched cruise-missile strikes against facilities in Afghanistan and Sudan in response to the 1998 bombing of U.S. embassies in Tanzania and Kenya. But they had no discernible impact on al Qa'ida. The use of massive military power against terrorist groups also runs a significant risk of turning the population against the government.

Military force has been most effective against large insurgent groups. Even in these cases, however, police and intelligence services were critical. An insurgency is an armed conflict that pits the government and national army of an internationally recognized state against one or more armed opposition groups able to mount effective resis-

---

[68] Galula (2005, p. 72).

[69] See, for example, Edward Luttwak, *Strategy: The Logic of War and Peace*, Cambridge, Mass.: Belknap Press of Harvard University Press, 2001.

[70] Pillar (2001, pp. 97–110); Hoyt (2004).

tance against the state.[71] In 70 percent of the cases since 1968 in which military force was effective, it was against groups that had more than 100 members. For example, military force was successful in defeating the Revolutionary United Front (RUF) in Sierra Leone. Britain was especially helpful in deploying approximately 4,500 soldiers and an aircraft carrier to Sierra Leone in 2000, which were pivotal in routing RUF forces and stabilizing the country.[72] As one 19-year-old student in Sierra Leone noted, "We love the British soldiers—they are showing some military guarantee. They are well equipped. They are not afraid like the U.N., like the Kenyans and Zambians who just gave up their arms and were taken hostage. They are not here to take any rubbish."[73] In Uruguay, the military unleashed a bloody campaign of mass arrests and selected disappearances against the Tupamaros, dispersing those guerrillas who were not killed or arrested. By 1972, the Tupamaros had been severely weakened, and its principal leaders were imprisoned. Even during insurgencies, however, there are limits to how military force can be used. As David Galula argued, "[C]onventional operations by themselves have at best no more effect than a fly swatter. Some guerrillas are bound to be caught, but new recruits will replace them as fast as they are lost."[74]

Victory

Since 1968, 10 percent of the terrorist groups that ended did so because they had achieved their goals. As Nobel Laureate in Economics Thomas Schelling wrote, terrorist groups may accomplish "intermediate means toward political objectives . . . but with a few exceptions it is hard to see

---

[71] On definitions of insurgency and civil war, see, for example, Michael W. Doyle and Nicholas Sambanis, *Making War and Building Peace: United Nations Peace Operations*, Princeton, N.J.: Princeton University Press, 2006, p. 31; Stathis N. Kalyvas, *The Logic of Violence in Civil War*, Cambridge and New York: Cambridge University Press, 2006, p. 5; and Fearon and Laitin (2003).

[72] See, for example, Richard Connaughton, "Operation 'Barass,'" *Small Wars and Insurgencies*, Vol. 12, No. 2, Summer 2001, pp. 110–119; and Doyle and Sambanis (2006, p. 318).

[73] Norimitsu Onishi, "British Plans to Leave Sierra Leone Prompt Worry," *New York Times*, June 7, 2000, p. A14.

[74] Galula (2005, p. 72).

that the attention and publicity have been of much value except as ends in themselves."[75] When they have achieved victory, it has usually been because they had narrow goals, such as policy or territorial change. No terrorist group that sought empire or social revolution has achieved victory since 1968. Perhaps more importantly, in most cases, terrorism had little or nothing to do with the outcome. There is rarely a causal link between the use of terrorism and the achievement of its goals.

One of the most significant exceptions is the African National Congress (ANC). The ANC was created in 1912 and turned to terrorism in the 1960s. ANC leader Nelson Mandela, imprisoned for terrorist acts from 1964 to 1990, was elected South Africa's first president following the end of apartheid. The last ANC attack occurred in 1989, and the organization became a legal political actor in 1990. As Mandela remarked at his 1964 trial, "without violence there would be no way open to the African people to succeed in their struggle against the principle of white supremacy. All lawful modes of expressing opposition to this principle had been closed by legislation, and we were placed in a position in which we had either to accept a permanent state of inferiority, or to defy the Government."[76] There are also a number of historical examples of successes. For example, the FLN against the French in Algeria (1954–1962) and the Irgun and Lohamei Herut Israel (Lehi) against the British during the Palestine Mandate (1937–1948) were successful in achieving their objectives. In both cases, terrorism played an important role in change, but it was not sufficient in Algerian independence or the creation of the state of Israel.[77]

Terrorism's ineffectiveness cuts against the prevailing view among much of the literature on terrorism, which assumes that terrorism is an

---

[75] Thomas Schelling, "What Purposes Can 'International Terrorism' Serve?" in R. G. Frey and Christopher W. Morris, eds., *Violence, Terrorism, and Justice*, Cambridge and New York: Cambridge University Press, 1991, pp. 18–32, p. 20.

[76] Nelson Mandela, "An Ideal for Which I Am Prepared to Die: Nelson Mandela April 20 1964," *Guardian*, April 23, 2007.

[77] On the FLN and Irgun cases as successful cases of terrorism, see Crenshaw (1996, pp. 260–261) and Hoffman (2006, pp. 43–62).

effective coercive strategy.[78] As Rohan Gunaratna concluded, "Theorists have long argued that terrorism does not work, but the sad fact is that it does."[79] Alan Dershowitz argued, for example, terrorism "works" and is "an entirely rational choice to achieve a political objective."[80] Andrew Kydd and Barbara Walter concluded that terrorist groups are "surprisingly successful in their aims."[81] In addition, the finding that the success rate of terrorist groups corresponded to the breadth of their goals also cuts against explanations of why terrorists do not achieve victory. Some argue, for instance, that the key variable for terrorist success is a tactical one: target selection. Groups whose attacks on civilian targets outnumber attacks on military targets systematically fail to achieve their policy objectives, in part because they fail to communicate their policy objectives well. Even when a terrorist group has limited objectives, target countries infer from attacks on their civilians that the group wants to destroy the country's values, society, or both. "In short," one study concluded, "target countries view the negative consequences of terrorist attacks on their societies and political systems as evidence that the terrorists want them destroyed."[82]

Insights from the Statistical Analysis

The database of 648 terrorist groups forms the basis for the statistical results that follow. For each terrorist group, our data include the country to which it is best associated, the year it started, the year it ended, its primary classification, its goals, its peak size, and the income and political freedom of the associated country. Finally, we recorded whether the group is active. If it is not active, we examine how it ended:

---

[78] Sprinzak (2000, p. 68); Pape (2005); Dershowitz (2002); Lake (2002); Kydd and Walter (2006); Trager and Zagorcheva (2005–2006).

[79] Rohan Gunaratna, *Inside Al Qaeda: Global Network of Terror*, New York: Columbia University Press, 2002, p. 321.

[80] Dershowitz (2002, p. 86).

[81] Kydd and Walter (2002, p. 264).

[82] Abrahms (2006, p. 59).

via victory, via defeat by policing or military forces, or through politics or splintering. As noted, the primary finding of this analysis is that policing and politics were far more likely to bring terrorist groups to an end than was military force or victory. As Figure 2.4 highlights, of the 648 groups we examined, 244 are alive and 136 splintered (thereby ending the group but not ending the terrorism), leaving 268 that came to an end in ways that eliminated their contribution to terrorism.

Of the 268 complete endings, the major share came from politics (43 percent) and policing (40 percent), with victory (10 percent) and military force (7 percent) far more the exception rather than the rule. Digging behind these numbers suggests, however, that terrorist groups differ in complex and revealing ways: (1) Religious terrorist groups are different from those otherwise classified, (2) terrorist groups in high-income countries are different from those in less-developed countries, and (3) large terrorist groups are different from small terrorist groups. In

Figure 2.4
Breakdown of the 648 Groups

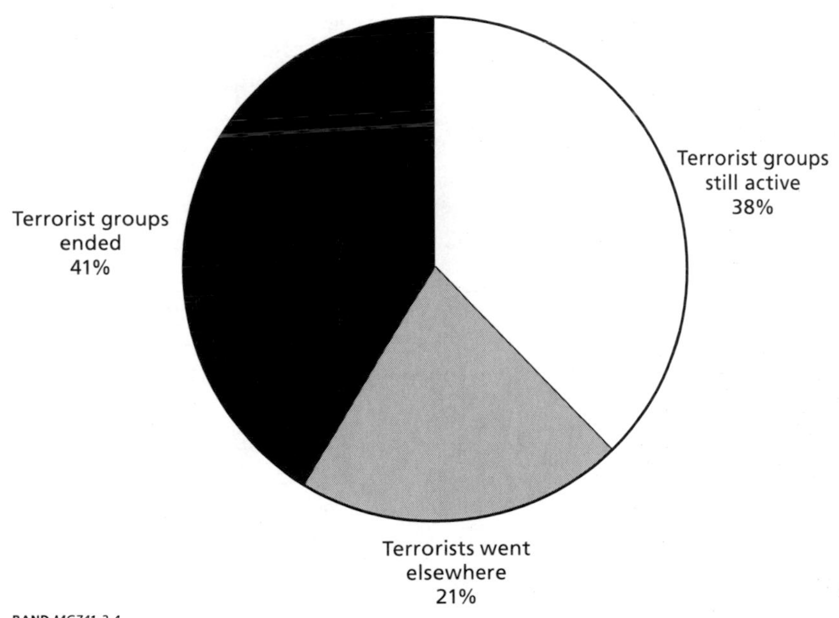

addition, terrorist groups capable of conducting an insurgency require different treatment from that given to terrorist groups that do not rise to the level of insurgents.

The most salient fact about religious terrorist groups is how hard they are to eliminate. As noted, 62 percent of all terrorist groups have ended, but only 32 percent of religious terrorist groups have ended. Does this result primarily from the fact that religious terrorist groups (median start year 1997) are more recent than the other groups (median start year 1988)? Our analysis suggests that it does not. As Figure 2.5 shows, the survivability of religious groups is substantially higher within *each* cohort. Of the 45 religious terrorist groups that *did* end, most (26) splintered. None achieved its aims. Of those that ended under pressure from authorities, most (13) fell to policing, while only

Figure 2.5
Survivability of Religious Groups

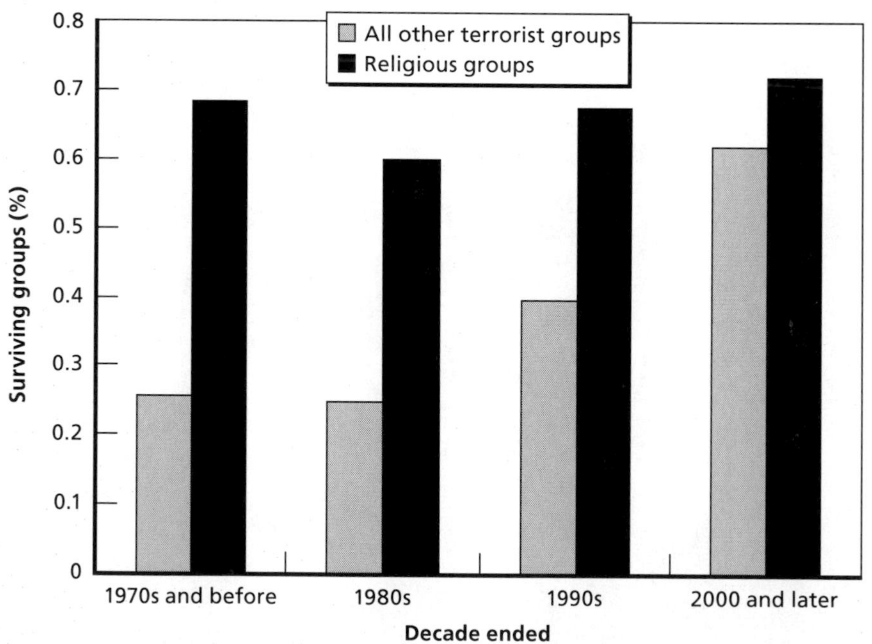

RAND *MG741-2.5*

three succumbed to military force. A mere three more abandoned terrorism for a political role.

Terrorist groups that operate in upper-income countries tend to be smaller than those that operate in less affluent countries, as noted in Figure 2.6. They also tend to be composed differently. Terrorist groups from upper-income countries are much more likely to be left-wing or nationalist and much less likely to be motivated by religious passions.

Figure 2.7 shows that terrorist groups in upper-income countries tend to end in a different manner. Far more of them tend to end than do groups in less developed nations. Another important difference is that policing is implicated in most of the terrorist-group ending in upper-income countries. Indeed, among the 96 terrorist groups that have ended and taken their terrorists with them, so to speak, a full 92 of them ended either via policing or politics.

Figure 2.6
Income Level of Groups

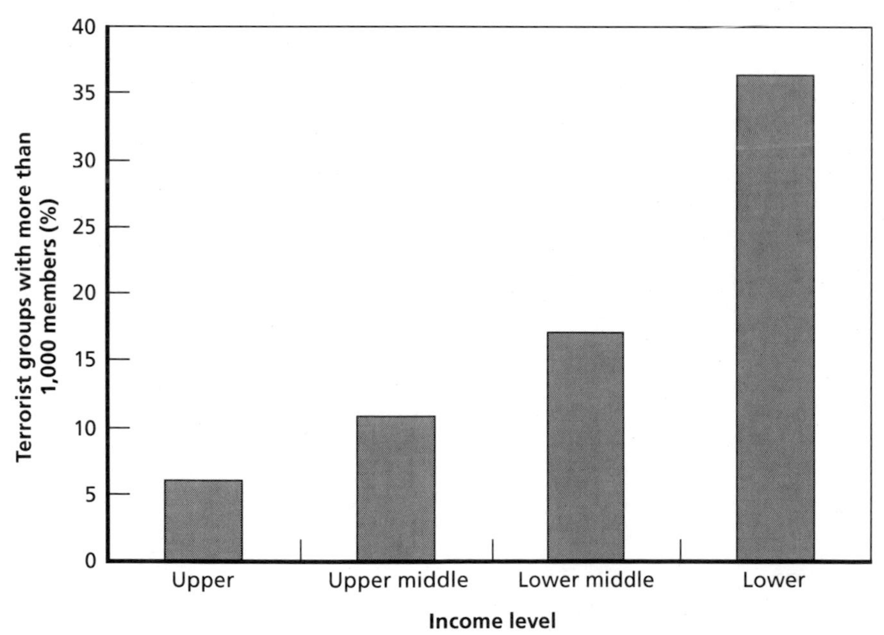

RAND *MG741-2.6*

Figure 2.7
Income Level and End of Groups

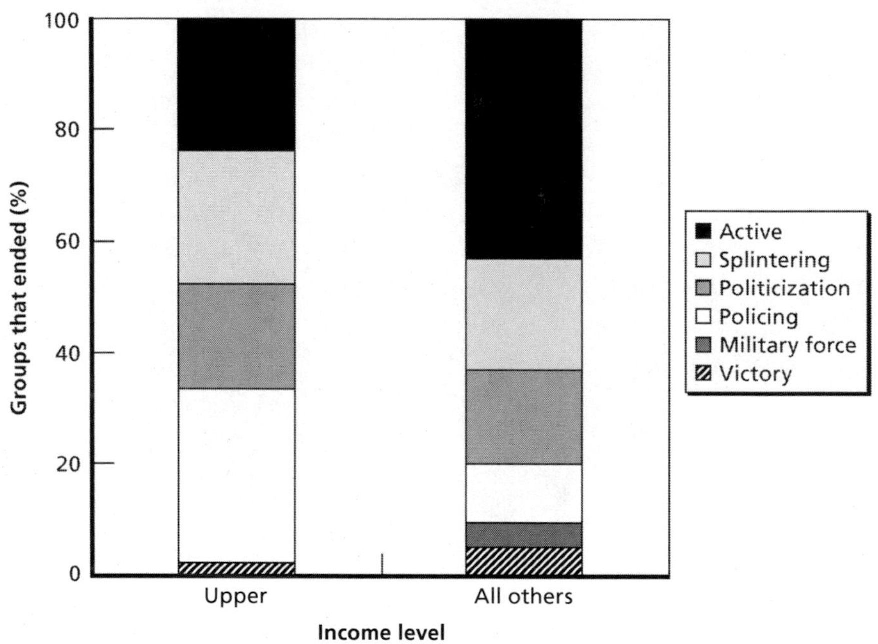

RAND *MG741-2.7*

Statistics also suggest that there are large differences between smaller and larger terrorist groups. In Figure 2.8, very large terrorist groups are those with more than 10,000 members (at their peak or to date); large groups have between 1,000 and 10,000 members; medium-sized groups have between 100 and 1,000 members; and small groups have fewer than 100. As shown in Figure 2.8, size is negatively correlated with host-country income. Only two of the 30 very large terrorist groups operate in high-income countries. By contrast, nearly 40 percent of the small groups can be found there. There are two possible explanations. On the one hand, rich countries tend to be democratic and politically stable and therefore avoid having disaffected citizens in numbers sufficient to form such large groups. On the other hand, small terrorist groups in high-income and media-saturated countries may be more likely to show up in the terrorism database (which is built

Figure 2.8
Size and Income Level of Groups

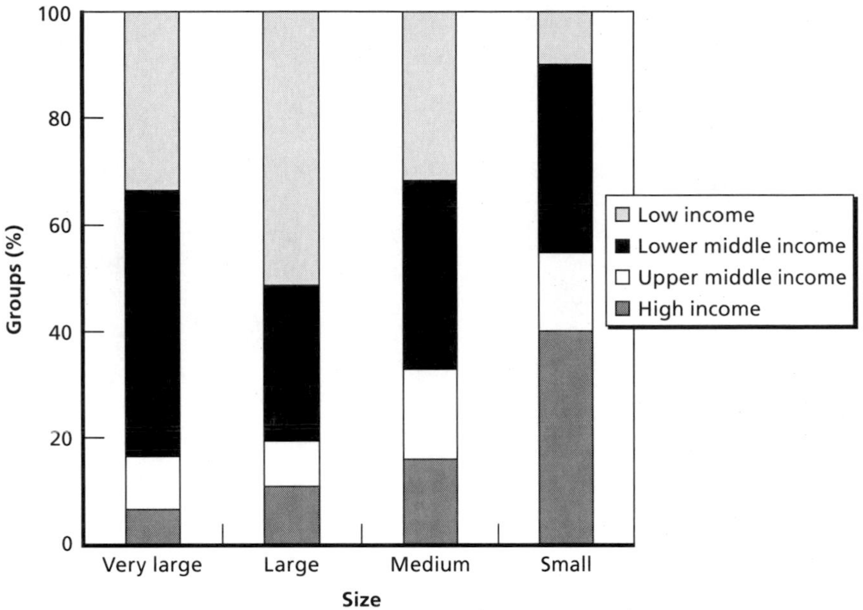

RAND *MG741-2.8*

from media reports) than would similar groups in countries with less income.

We next characterize groups by type, as illustrated in Figure 2.9. The influence of size, while significant, is not as dominant as in the last chart. Very large groups are likely to be of the left-wing or nationalist type. Medium-sized groups, by contrast, are more apt to be religious. Otherwise, the differences are modest.

Size is more clearly correlated with goals. Figure 2.10 shows that roughly half of the very large groups seek regime change. Conversely, although almost 30 percent of the small groups seek only policy change, fewer than 10 percent of groups with more than 100 members want something so limited. Apart from these two goals, the correlation between size and goals is relatively weak.

Size, however, is a significant determinant of a group's fate. Figure 2.11 indicates that big groups have been victorious more than 25 percent of the time, while victory is rare when a group is smaller than

Figure 2.9
Size and Type of Groups

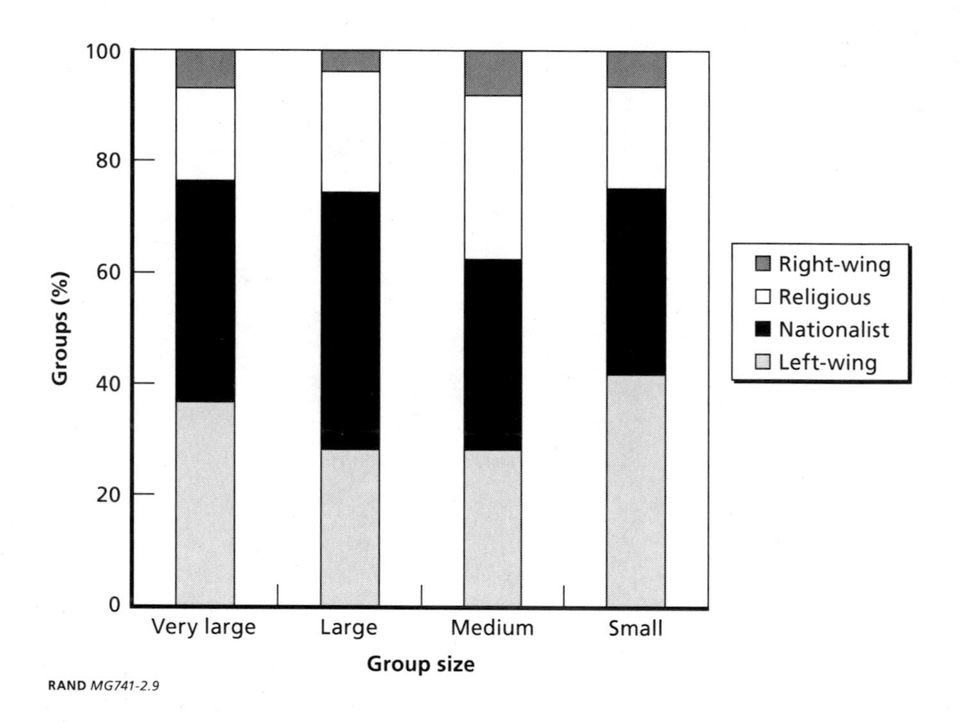

1,000 members. A terrorist group often has to become large to win. The inability to grow, conversely, is a harbinger of defeat. Splintering (or absorption into other groups) is a bane of small groups, again for straightforward reasons: Larger groups are ones that have stood the test of time and can stand on their own. The tendency of large and very large groups to end by becoming part of the political process is striking. Policing is quite effective for groups that have yet to reach 1,000 members and almost never effective when groups are larger. Conversely, military force has claimed roughly 10 percent of the large and very large groups, but it is rarely used against medium-sized and small groups.

Our statistical analysis, which is included in Appendix A, showed that there is no correlation between the duration of terrorist groups and ideological motivation, economic conditions, regime type, or the breadth of terrorist goals. The best single-factor correlation was with

Figure 2.10
Size and Goals of Groups

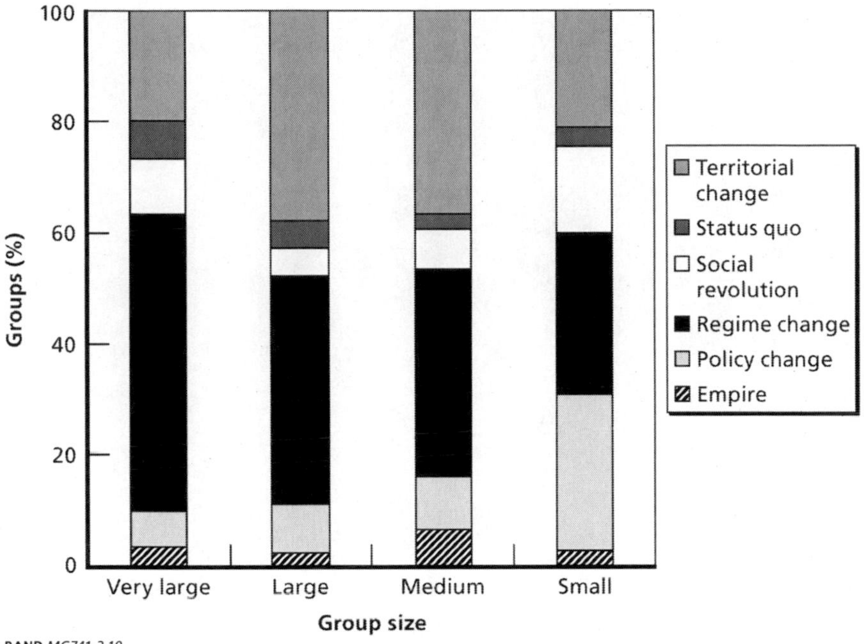

RAND *MG741-2.10*

peak size, which alone explained 32 percent of the variance in outcome scores (i.e., $r^2 = 0.32$). No other single variable on its own explained more than 4 percent of the variance. Remarkably, the addition of all other explanatory variables together improved the explanatory power only from 32 to 34 percent—or hardly at all. On the surface, there appears to be weak support for the argument that a terrorist group's size may affect its duration: Larger groups tend to last longer than smaller groups. However, size may be endogenous to duration. That is, groups that last longer may be larger simply because they have more time to recruit new members. In this case, duration would cause size, rather than the other way around.

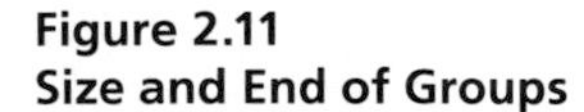

Figure 2.11
Size and End of Groups

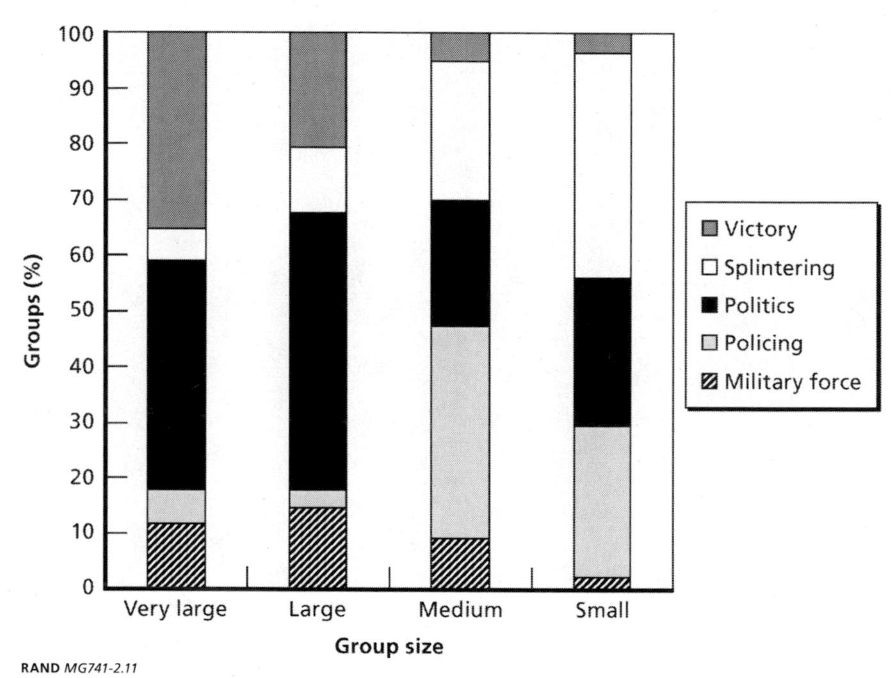

RAND *MG741-2.11*

Conclusions

How do terrorist groups end? The evidence presented in this chapter suggests that most terrorist groups do not end through military force and that groups rarely achieve victory. Rather, groups end because they adopt nonviolent tactics and join the political process or because local law-enforcement agencies arrest or kill key members of the group. In cases in which a terrorist group has minimal goals, such as policy change, governments may be able to reach a negotiated settlement with the group. But where there is no bargaining room to allow terrorist groups back into the political arena, policing is most effective in defeating terrorist groups. Police have a permanent presence in cities, towns, and villages; a better understanding of local communities than other security forces; and better intelligence. This enables them to be best suited to understand and penetrate terrorist networks. However, the causal process is not entirely clear. How does policing work in practice? And how, specifically, do police infiltrate and destroy terrorist groups? To examine these questions, the next several chapters examine case studies.[83] As Alexander George and Timothy McKeown argued, case studies are useful in uncovering

> what stimuli the actors attend to; the decision process that makes use of these stimuli to arrive at decisions; the actual behavior that then occurs; the effect of various institutional arrangements on attention, processing, and behavior; and the effect of other variables of interest on attention, processing, and behavior.[84]

---

[83] On the costs and benefits of comparative case studies, see Collier (1991); Ragin (1981); Tilly (1997); Skocpol and Somers (1980); and Van Evera (1997, pp. 49–88).

[84] George and McKeown (1985, p. 35).

CHAPTER THREE

Policing and Japan's Aum Shinrikyo

At 7:45 a.m. on March 20, 1995, five members of the terrorist group Aum Shinrikyo boarded trains at different ends of Tokyo's sprawling subway system. They coordinated their attack so that the trains would converge a half-hour later at a single central stop: the Kasumigaseki station in the heart of Tokyo's government district. The station was strategically situated several blocks from the Japanese parliament building, government agencies, and the Imperial Palace. The Aum Shinrikyo terrorists who carried out the attacks included a young graduate student in physics at Tokyo University, a former cardiovascular surgeon who had graduated from Keio University, a former physics student from Waseda University, and an electronics engineer. Each member carried an umbrella with a sharpened tip and held a loosely wrapped newspaper in his arms. Inside the newspaper was a plastic sack containing sarin. As the trains converged on the Kasumigaseki station, the Aum Shinrikyo members put their newspapers on the floor of the train and pierced the plastic sack with the sharpened end of the umbrella. They quickly exited the trains and left behind leaking plastic bags with sarin gas, which permeated the subway trains. The attack killed 12 people and injured more than 5,000 others.[1] Over the subsequent several years, Japanese police and intelligence officials conducted one of the largest manhunts in the country's history. By 1997, Aum Shinrikyo had been eradicated as a terrorist organization. Most of its leadership structure

---

[1] Mark Juergensmeyer, *Terror in the Mind of God: The Global Rise of Religious Violence*, Berkeley, Calif.: University of California Press, 2000, pp. 103–104.

was in jail, and the organization was bankrupt. It eventually changed its name to Aleph but ceased to be involved in terrorism.

How did Aum Shinrikyo end as a terrorist organization? This chapter argues that efforts by Japanese police and intelligence services were fundamental to ending Aum Shinrikyo as a terrorist organization. They conducted widespread surveillance and penetration of Aum Shinrikyo, made hundreds of arrests, and adopted a range of legal measures that crippled the organization's financial base. They also discredited the group's ideology, leading to a mass exodus of supporters. In short, policing was an effective strategy against Aum Shinrikyo.

In examining Aum Shinrikyo, this chapter is organized into three sections. The first examines Aum Shinrikyo's evolution and goals. The second section examines how policing was effective in ending Aum Shinrikyo as a terrorist organization. And the third section offers conclusions on policing and the end of terrorist groups.

Aum Shinrikyo's Ideology

Shōkō Asahara, a blind former yoga teacher, founded Aum Shinrikyo in the late 1980s. The group's ideology was grounded largely in Buddhism. But it had a strong mixture of assorted eastern and western mystical beliefs and drew from the works of the 16th-century French apothecary Nostradamus. Asahara preached that there were a number of levels of consciousness that a member could reach through his teachings. But only one person, Asahara himself, had attained the highest level of consciousness and existed in the state of Nirvana. At the core of Asahara's worldview was a great cloud casting its shadow over the future: the specter of a world catastrophe unparalleled in human history. Although World War II had been disastrous to Japan, especially with the nuclear weapons dropped on Hiroshima and Nagasaki, it paled in comparison with what Asahara described as World War III. The term that Asahara chose for this cataclysmic event was *Armageddon*.

In 1989, for example, Asahara published a major religious treatise on Armageddon called *The Destruction of the World*. In it, he described a worldwide calamity based on a purported war between Japan and the

United States that would start sometime in 1997.[2] He based his predictions on the prophecies of Nostradamus, the Book of Revelation in the Christian Bible, Buddhist scripture, and other personal revelations. In 1993, Asahara publicly reiterated his predictions of Armageddon. In a book titled *Shivering Predictions by Shoko Asahara*, he stated that,

> From now until the year 2000, a series of violent phenomena filled with fear that are too difficult to describe will occur. Japan will turn into waste land as a result of a nuclear . . . attack. This will occur from 1996 through January 1998. An alliance centering on the United States will attack Japan. In large cities in Japan, only one-tenth of the population will be able to survive. Nine out of ten people will die.[3]

When Armageddon came, the evil forces would attack with the most vicious weapons: "Radioactivity and other bad circumstances—poison gas, epidemics, food shortages—will occur," he claimed. The only people who would survive were those "with great karma" and those who had the defensive protection of the Aum Shinrikyo organization. "They will survive," Asahara said, "and create a new and transcendent world."[4]

Most of Asahara's prophecies predicted that Armageddon would occur in 1997 or 1998. But documents that the Japanese police seized after the Tokyo attack indicated that, sometime in 1994, Asahara had

---

[2] Indeed, Asahara repeatedly alleged that the U.S. military was attacking Aum and its facilities with chemical weapons. See, for example, John Parachini, "Aum Shinrikyo," in Brian A. Jackson, John C. Baker, Peter Chalk, Kim Cragin, John V. Parachini, and Horacio R. Trujillo, Aptitude for Destruction, Vol. 2: Case Studies of Organizational Learning in Five Terrorist Groups, Santa Monica, Calif.: RAND Corporation, MG-332-NIJ, 2005, pp. 11–36, p. 14.

[3] U.S. Senate Committee on Governmental Affairs Permanent Subcommittee on Investigations, *Staff Statement, U.S. Senate Permanent Subcommittee on Investigations (Minority Staff): Hearings on Global Proliferation of Weapons of Mass Destruction: A Case Study on the Aum Shinrikyo*, Washington, D.C., 1995.

[4] Shōkō Asahara, *Disaster Approaches the Land of the Rising Sun: Shoko Asahara's Apocalyptic Predictions*, Shizuoka, Japan: Aum Publishing Co., 1995, pp. 135–136.

moved the date for this cataclysmic event up to 1995.[5] It appears that Aum may have decided to speed things up by instigating the predicted war between Japan and the United States in 1995. The new timetable for Armageddon appears to have coincided with public statements by Asahara that he and his people were already the victims of gas attacks by Japanese and U.S. military aircraft. In a public sermon delivered by Asahara at his Tokyo headquarters on April 27, 1994, for example, he claimed,

> With the poison gas attacks that have continued since 1988, we are sprayed by helicopters and other aircraft wherever we go. . . . The use of poison gases such as sarin were clearly indicated. The hour of my death has been foretold. The gas phenomenon has already happened. Perhaps the nuclear bomb will come next.[6]

The date of this speech is significant, since it predated by two months the June 27, 1994, Aum Shinrikyo sarin-gas attack in Matsumoto, Japan. The attack left seven dead and hospitalized 500 others and was soon followed by the even more lethal sarin attack in Tokyo in March 1995 that killed 12 and injured 5,000 others.

The End of Aum Shinrikyo

How did Aum Shinrikyo end as a terrorist organization? Following the Tokyo subway attack, Japan effectively adopted a policing strategy that included the collection and analysis of intelligence, arrest of key leaders, and adoption of a range of legal measures that crippled the organization. The result was the end of Aum Shinrikyo as a terrorist organization, though the organization continued to operate as a nonviolent cult.

---

[5] U.S. Senate Committee on Governmental Affairs Permanent Subcommittee on Investigations (1995).

[6] U.S. Senate Committee on Governmental Affairs Permanent Subcommittee on Investigations (1995).

Intelligence Collection

Until the Tokyo attack, Aum Shinrikyo had benefited from a loosely organized and relatively weak Japanese intelligence apparatus. This was largely a legacy of Japan's World War II military government. To some extent, this resulted from limitations in the Japanese constitution. For instance, there were legal bans on the police use of preventive surveillance techniques.[7] Shielded by limited governmental powers, Aum Shinrikyo was able to accumulate extensive stockpiles of cash and dangerous chemical and biological weapons without raising police suspicion. In addition, Japan's intelligence service, the Public Security Intelligence Agency, was set up at the height of the Korean War in 1952 to monitor and target communist groups in Japan. But since the end of the Cold War and the collapse of communism, the absence of clear enemy forces undermined its position and status.[8]

However, this changed in the aftermath of the Tokyo subway attack. Japanese police and intelligence agencies began a major intelligence-collection and investigation effort. The Public Security Intelligence Agency placed Aum Shinrikyo under surveillance in accordance with the Act Pertaining to Control of Organizations That Commit Indiscriminate Murder, which was enacted after the Tokyo attack.[9] Aum Shinrikyo members fled across the country, concealing various pieces of evidence. The Japanese Criminal Affairs Bureau and the Security Bureau jointly established the Task Force for Toxic Agent Attack in Tokyo Subways at the National Police Agency. The investigational headquarters for the crime, led by the director of criminal investigation, was established at the Tokyo Metropolitan Police Department's Tsukiji station so that police could pursue an initial

---

[7] Robyn Pangi, "Consequence Management in the 1995 Sarin Attacks on the Japanese Subway System," *Studies in Conflict and Terrorism*, Vol. 25, No. 6, November–December 2002, pp. 421–448, p. 422.

[8] Ian Reader, "Spectres and Shadows: Aum Shinrikyo and the Road to Megiddo," *Terrorism and Political Violence*, Volume 14, Number 1, Spring 2002, pp. 145–186; Peter J. Katzenstein, "Same War: Different Views: Germany, Japan, and Counterterrorism," *International Organization*, Vol. 57, No. 4, Autumn 2003, pp. 731–760.

[9] Mizukoshi Hideaki, "Terrorists, Terrorism, and Japan's Counter-Terrorism Policy," *Gaiko Forum*, Vol. 3, No. 2, Summer 2003, pp. 53–63.

investigation, rescue victims, and secure witnesses. The police established a countermeasure group for pursuit and investigation at the National Police Agency Criminal Investigation Bureau and an investigation headquarters at each prefectural police station. The police also distributed approximately 1.6 million posters and brochures to reach out to the local population to aid in tracking down those suspects still at large.[10]

After the Tokyo attack, police seized as evidence the plastic bags and remaining liquid left in the subway trains and confirmed that the substance was sarin using a gas chromatography–mass spectrometry test. On March 22, 1995, the police raided 25 of Aum Shinrikyo's offices, compounds, and complexes throughout Japan. This included the Aum Shinrikyo facility in Kamikuishiki, where police wore gas masks and carried canaries.[11] Approximately 2,500 police department staff members were involved in the investigation.[12] Department staff worked in dangerous environments where poisonous chemical substances, such as phosphorus trichloride, were stockpiled and wore chemical protective suits while searching for and confiscating evidence. At Aum facilities, police seized large quantities of illegal drugs and chemicals for the production of sarin and other dangerous substances.[13] Indeed, at Aum's facility at Mount Fuji, the police and intelligence services collected a mammoth stockpile of chemicals, such as sodium cyanide, hydrochloric acid, chloroform, phenylacetonitrile for stimulant production, glycerol for explosives, huge amounts of peptone for cultivating bacteria, sodium fluoride, and 500 drums of phosphorus trichloride. Japanese officials estimated that Aum's stockpile included more than 200 different kinds of chemicals.[14]

---

[10] National Police Agency, *1996 Police White Paper*, Tokyo, 1996, p. 20.

[11] David E. Kaplan and Andrew Marshall, *The Cult at the End of the World: The Terrifying Story of the Aum Doomsday Cult, from the Subways of Tokyo to the Nuclear Arsenals of Russia*, New York: Crown Publishers, 1996, p. 255; Pangi (2002, p. 438).

[12] NPA (1996, p. 20).

[13] Robert Jay Lifton, *Destroying the World to Save It: Aum Shinrikyo, Apocalyptic Violence, and the New Global Terrorism*, New York: Henry Holt and Co., 1999, pp. 56–57.

[14] Kaplan and Marshall (1996, p. 257).

One of the most challenging aspects of evidence collection was acquiring expertise on biological and chemical agents. As one police report noted, the Japanese police "lacked information regarding the characteristics, toxicity, manufacturing process, and raw materials of [sarin], as well as of other poisonous gases such as VX."[15] The police hurried to acquire information on sarin and other chemical and biological agents, their manufacturing process, and the buyers of the components necessary for production. The Japanese military—the Japan Self-Defense Forces (JSDF)—sent chemical-warfare experts to assist the police. The National Police Agency and JSDF subsequently established a joint police and army investigative unit.[16] However, the lack of knowledge and information on sarin and other agents made the investigation more difficult. Indeed, although many experts cooperated in helping the police learn more about the production process for sarin, the investigation was hindered, since few of these experts had actually produced sarin, and various methods of production were found to be theoretically plausible.[17]

These deficiencies led the police to restructure the National Police Agency to strengthen its scientific-investigation, information-collection, and data-analysis systems. To equip investigators with the technical knowledge and skills necessary for scientific investigations, the National Police Agency, regional police bureau, and prefectural police stations promoted scientific studies—including the study of chemistry and physics. The National Police Agency reinforced its data-gathering and -analysis system through the creation of a special organized-crime group. The National Police Agency also created the Police Policy Research Center to analyze drastic transitional social phenomena and group and individual attitudes and behaviors from a public-security perspective.[18] In January 2000, Aum Shinrikyo was placed under surveillance for three years under an anti-Aum law that required the group

[15] NPA (1996, p. 25).

[16] Pangi (2002, pp. 427, 429).

[17] NPA (1996, p. 25).

[18] NPA (1996, pp. 27–28).

to submit a list of members and details of assets to the authorities. In January 2003 and 2006, Japan's Public Security Intelligence Agency received permission to extend the surveillance for another three years to monitor the group's activities.

Arrest of Key Leaders

Between March 22 and May 16, the police arrested more than 200 Aum Shinrikyo members.[19] They had questioned several Aum Shinrikyo members taken into custody shortly after the Tokyo sarin attacks who had identified a number of individuals involved in the Tokyo attack and key leaders. On May 16, 1995, investigators searching through an Aum facility in Kamikuishiki found Shōkō Asahara and arrested him. On the same day, the police arrested 41 Aum Shinrikyo members on murder and murder-accomplice charges and designated four other Aum Shinrikyo members as National Police Agency most-wanted suspects.[20] Asahara was put on trial. He pleaded not guilty to all charges, claiming that his followers acted without his knowledge. Nevertheless, he was sentenced to death. Some Aum leaders received death sentences, while others received life sentences. Many members sought appeals, but Japanese courts rejected most of them. On September 15, 2006, Asahara lost his final appeal against the death penalty.

Indeed, Aum members were indicted on various charges, including the subway gas attack, illegal production of drugs, and a range of violent acts. Interviews with a number of former Aum members demonstrated that illicit narcotics, such as LSD, were integral parts of some Aum Shinrikyo ceremonies.[21] Aum Shinrikyo was involved in a number of violent crimes other than the 1995 sarin-gas attack: the November 1989 kidnapping and murder of Tsutsumi Sakamoto, a lawyer representing concerned parents of Aum members, along with his wife and son; the February 1994 lynching of Kotara Ochida, an unco-

[19] Mark Mullins, "The Legal and Political Fallout of the 'Aum Affair,'" in Robert Kisala and Mark Mullins, eds., *Religion and Social Crisis in Japan: Understanding Japanese Society Through the Aum Affair*, New York: Palgrave, 2001, pp. 71–72.

[20] NPA (1996, p. 20).

[21] See, for example, Juergensmeyer (2000, pp. 103–118) and Lifton (1999).

operative Aum member; the June 1994 sarin-gas attack in Masumoto that killed seven and injured more than 200 people; and the February 1995 kidnapping and murder of Kiyoshi Kariya, a Tokyo notary.[22] By mid-1996, more than 400 cult members had been arrested. Most of the arrests were based not on a connection to terrorism or the sarin attack but on kidnapping or drug charges.[23] For example, Tomomitsu Niimi (Aum's minister of internal affairs) and Ikuo Hayashi (who ran cult clinics and presided over human experimentation) were arrested for confining followers; Kiyohide Hayakawa (Aum's minister of construction) was arrested for trespassing; Fumihiro Joyu (who ran the Moscow office and then became Aum's spokesperson) was arrested for perjury; and Yoshinobu Aoyama (Aum's attorney) was arrested for libel.[24]

As information regarding Aum Shinrikyo's involvement in the Tokyo attacks began to surface, residents who lived near Aum Shinrikyo facilities began to feel insecure and called police for protection and assistance. In an effort to allay these concerns, Japanese police increased patrols on foot and by car in these areas. For example, in Fujigamine, where there were several Aum Shinrikyo facilities, the police opened a temporary regional-security center in July 1995 to provide residents with patrol and consultation services. To allay concerns among residents in the Hitoana area, where Aum Shinrikyo's Mount Fuji headquarters was located, the police established a temporary station in an open lot in front of the Aum Shinrikyo facility.[25] In some cases, the local population took independent action. In the town of Kitamimaki, for example, farmers dug a moat around an Aum Shinrikyo facility and staffed guard towers for six months.[26] In addition, police offered consultation services to members of Aum Shinrikyo and their families who wanted to leave the cult and return to society. The police, in

---

[22] Mullins (2001, p. 71).

[23] Mark Mullins, "The Political and Legal Response to Aum-Related Violence in Japan: A Review Article," *Japan Christian Review*, Vol. 63, 1997, pp. 37–46; Pangi (2002, p. 438).

[24] Kaplan and Marshall (1996, p. 269).

[25] NPA (1996, p. 24).

[26] George Wehrfritz, "Crushing the Cult of Doom," *Newsweek*, November 22, 1999, p. 44.

cooperation with child-care clinics and schools, also provided advice to youth members who left Aum Shinrikyo. To develop a comprehensive system for dealing with Aum Shinrikyo returnees, the National Police Agency created a special Adjuster for Return Measures for Members of Aum Shinrikyo and the Concerned position in December 1995 to deal with these cases.[27]

There were also some arrests made abroad. In July 2000, for example, Russian police arrested Dmitri Sigachev, ex-KGB and former Aum Shinrikyo member, and four other former Russian Aum members, for stockpiling weapons in preparation for attacking Japanese cities in a bid to free Asahara. Indeed, Aum's most successful recruitment spree had occurred in Russia. In 1992, Asahara and an entourage of several hundred Japanese members made a salvation tour to Russia. In the 18 months following Asahara's tour, the group's Russian membership surged. Over a three-year period, the group attracted between 30,000 and 40,000 Russian followers.[28] Aum used its tremendous wealth to gain access to high-ranking officials, including the Russian vice president, the head of the Russian parliament, and the secretary of the security council. Some reports estimated that Aum paid "$12 million in payoffs to well-placed officials" in Russia.[29] Russian press reports claimed that Aum's overall investment in Russia "amounted to some $50 million."[30]

Legal Measures

Prior to the 1995 attacks, Aum Shinrikyo was able to escape police scrutiny partially because of legal barriers against religious prosecution. In 1951, the Japanese government passed the Religious Corporation Law (*Shukyo Hojin Ho*). It strengthened constitutionally guaranteed religious freedoms by relieving any organization that could be identified as "religious" of tax obligations and by providing these groups with

---

[27] NPA (1996, p. 24).

[28] Parachini (2005, p. 29).

[29] Kyle B. Olson, "Aum Shinrikyo: Once and Future Threat?" *Emerging Infectious Diseases*, Vol. 5, No. 4, July–August 1999, p. 515; Parachini (2005, p. 29).

[30] Kaplan and Marshall (1996, p. 206).

unusually strong protection from government intrusion.[31] This law was a reaction to the harsh suppression of religious freedom by Japan's military government before World War II.[32] The Japanese police, like the government bureaucracy, exercised caution in handling complaints against official religious groups. Aum was permitted—like all other registered religious bodies—to own properties and buildings for worship and religious activities, as well as operate various business enterprises to support the organization's religious aims. A basic assumption behind the law was that registered religious organizations contribute to the public good (*kōeki*). Consequently, they were permitted to engage in economic activities to support their religious work and public-welfare activities (*kōeki jigyō*).[33]

Following the subway attacks, however, the Japanese government passed a series of laws that were retroactively applied to Aum Shinrikyo. For instance, the governor of Tokyo and public prosecutors at the Tokyo District Public Prosecutor's Office filed suit against Aum Shinrikyo in the Tokyo district court on June 30, calling for the group's forced breakup under the Religion Corporation Law. The Tokyo district court sided with the city and ordered Aum Shinrikyo's breakup on October 30.[34] According to the court, there was clear evidence that the leaders of Aum had been involved in numerous illegal and violent acts and no longer qualified as a religious corporation. The court concluded that it was an organization working *against* the public good and should not enjoy the favorable treatment accorded such corporations.[35] Aum Shinrikyo immediately appealed the decision through the Tokyo high court on November 2. However, its appeal was dismissed on December 19, and the cult's religious status was officially revoked.[36] Legally, this did not prevent Aum Shinrikyo members from practicing

---

[31] Pangi (2002, pp. 422–423).

[32] D. W. Bracket, *Holy Terror: Armageddon in Tokyo*, New York: Weatherhill, 1996, p. 13.

[33] Mullins (2001, p. 72).

[34] NPA (1996, p. 23); Pangi (2002, p. 438).

[35] Mullins (2001, p. 73).

[36] NPA (1996, p. 23); Pangi (2002, p. 438).

their faith or running affiliated business ventures. Rather, the cult lost the legal protections guaranteed to religious groups, including its tax-exempt status.

In addition, the Japanese Parliament passed the Law Related to the Prevention of Bodily Harm Caused by Sarin and Similar Substances. It prohibited the manufacture, possession, and use of sarin and similar substances. The laws guaranteeing religious freedom were modified after the sarin attack. On December 8, 1995, Japan's legislature (*Teikoku Gikai*) passed revisions to the Religious Corporations Law that increased the ability of authorities to monitor potentially dangerous religious organizations.[37] The government also began reviewing laws restricting police actions, antiterrorism policies, and consequence-management plans. In June 1996, the police law was revised "to enable prefectural police to extend their authority out of their border by their own judgment and responsibility in dealing with tranprefectural organized crimes."[38] A few years later, the Group Regulation Act of 1999 was passed to regulate groups that committed indiscriminate mass murder. The law did not directly refer to Aum Shinrikyo, but it did allow police to put the organization under surveillance for a maximum of three years.

Aum Shinrikyo quickly experienced significant financial trouble. Donations dropped off, and Aum faced numerous claims for compensation from victims and their relatives. The Japanese government, in conjunction with the victims of the Tokyo subway attack and their families, requested that Aum Shinrikyo's assets be frozen. These steps were intended to direct Aum Shinrikyo money into compensation funds for victims before the group's debts became too exorbitant. On December 12, under bankruptcy law, the Tokyo district court froze all Aum Shinrikyo assets and began administering the provisional attachment of the group's assets beginning on December 14. On March 28, 1996, the Tokyo district court made the adjudication of bankruptcy for Aum Shinrikyo.[39] The court also appointed an adjudicator to replace

---

37 Mullins (1997).

38 Pangi (2002, p. 439).

39 NPA (1996, p. 23).

Asahar as the group's legal representative and supervise the liquidation of Aum's assets (including its centers and commune), which were scattered across Japan and as far away as Russia and Germany.[40]

A number of other governments also took legal measures. For example, the Russian government banned Aum and ordered it to pay $4 million in compensation to the Youth Salvation Committee, a group led by Russian parents who had filed a civil suit claiming Aum had brainwashed and kidnapped dozens of followers.[41]

While several pieces of legislation were passed in direct response to the sarin attack, the government chose moderate language and limited the measures that restricted civil liberties. Once the government brought charges, it chose to treat the attacks as individual violations rather than bring charges under articles 77 (carrying out civil war) and 78 (preparing for civil war) of the Japanese constitution.[42] In May 1995, the Public Security Intelligence Agency initiated procedures to apply the Anti-Subversive Activities Law of 1952 to Aum Shinrikyo. This law would have prohibited Aum Shinrikyo from recruiting and fund-raising, training followers, or publishing materials promoting its beliefs. To invoke the law, the Public Security Intelligence Agency first had to exhaust all other legal remedies and then prove that the violence that the group committed was politically motivated and that there was a significant possibility that future acts of violence would be committed.[43] Indeed, although there was significant public support for the measures invoked, the majority of the population did not support invoking the Anti-Subversive Activities Law. Many viewed it as compromising freedom of speech and other civil liberties.[44] The Public Security Intelligence Agency completed its investigation and hearings on July 11, 1996, and formally asked the Public Security Commission

[40] Mullins (2001, p. 73).

[41] Kaplan and Marshall (1996, pp. 266–267).

[42] Katzenstein (2003, p. 746).

[43] Mullins (1997, pp. 37–46).

[44] Ian Reader, *Religious Violence in Contemporary Japan: The Case of Aum Shinrikyo*, Honolulu: University of Hawai'i Press, 2000, pp. 224–225; Pangi (2002, p. 439).

to invoke the law against Aum Shinkrikyo. But on January 31, 1997, the Public Security Commission decided that it would not invoke the Anti-Subversive Activities Law, on the grounds that there was insufficient evidence that Aum was capable of carrying out continued or repeated attacks in the future.[45]

After that, younger members of Aum Shinrikyo began to revive the group, though not as a terrorist organization. In a January 18, 2000, press conference, an Aum representative admitted that Shōkō Asahara and Aum had been involved in criminal activities and offered apologies to the victims and their surviving family members. Aum also changed its name to Aleph and stated that Asahara was no longer recognized as their leader. Furthermore, they noted that those doctrines legitimizing murder and violence had been eliminated. Fumihiro Joyu, the group's second-highest official, who served a three-year jail term for perjury, extended his apologies to victims of Aum Shinrikyo's crimes: "I'd like to give a deep apology to the victims and bereaved, and say that I feel personally responsible as one who belonged to the same religious group."[46] Nonetheless, Japanese police and intelligence agencies continued to monitor Aleph's activities. Between 2000 and 2005, for example, the Public Security Intelligence Agency conducted on-site inspections at 185 Aleph facilities in 19 prefectures across Japan.[47]

The Logic of Policing

A policing strategy was effective in ending Aum Shinrikyo as a terrorist organization. Japan's police and intelligence services responded

---

[45] Christopher W. Hughes, "The Reaction of the Police and Security Authorities to Aum Shinrikyo," in Robert Kisala and Mark Mullins, eds., *Religion and Social Crisis in Japan: Understanding Japanese Society Through the Aum Affair*, Basingstoke, Hampshire, UK, and New York: Palgrave, 2001, p. 65.

[46] "Aum Admits Matsumoto May Be Linked to Crimes," *Daily Yomiuri*, January 19, 2000, p. 1; Calvin Sims, "Poison Gas Group in Japan Distances Itself from Guru," *New York Times*, January 19, 2000, p. A6.

[47] Public Security Intelligence Agency, *The Review and Prospects of Internal and External Situations*, Tokyo, 2006.

to the Tokyo subway attack with a massive intelligence-collection and -analysis effort (including information gleaned from detained Aum Shinrikyo members). They developed a picture of Aum's organizational structure, tactics, strategies, production and use of chemical and biological agents, and involvement in terrorist attacks. They then arrested key members of the group, developed an information campaign to dissuade current and potential members from joining the organization, and tried to cripple the organization through legal measures. While Aum Shinrikyo continued to exist as a cult, changing its name to Aleph, it ceased to be involved in terrorism.

While successful against Aum Shinrikyo, policing is less likely to be effective against large insurgent groups that use terrorism. In these cases, military forces may be necessary, since they have greater firepower. What does the Aum Shinrikyo experience suggest about the end of terrorist groups more broadly? There are several issues worth considering.

First, Japan was a strong state that had, in Max Weber's words, a "monopoly of the legitimate use of physical force within a given territory."[48] This is not always the case. Adopting a policing strategy in weak states presents a significant challenge, though it does not preclude working with tribal groups or other substate actors. Second, Japan was committed to undermining Aum Shinrikyo. Unfortunately, a state may support terrorist groups operating on its soil, as Syria has historically done with Palestinian terrorist groups. Encouraging a policing strategy in these countries—and expecting local forces to undermine terrorist groups—may be wishful thinking. Third, Japan was democratic. Implementing a policing strategy in a nondemocratic state may be notably different because of disparate laws and norms of behavior, including the repressive nature of security forces. Fourth, while Aum Shinrikyo did possess some expertise in biological and chemical warfare and did conduct military training for some of its members, it was not capable of taking on Japanese security forces in pitched battles.

[48] Max Weber, "Politics as a Vocation," in Max Weber, Hans Heinrich Gerth, and C. Wright Mills, eds., *From Max Weber: Essays in Sociology*, New York: Oxford University Press, 1958, pp. 77–128, p. 78.

In some cases, terrorist groups may possess such significant military power—such as the Khmer Rouge, FMLN, Taliban, or Armed Islamic Group—that military forces are necessary to counter them. In sum, a policing strategy may be appropriate in many situations. But it may be less apropos in cases with a weak state, little political will to defeat terrorist organizations, or powerful insurgent groups.

Policing can be an effective strategy to penetrate terrorist organizations, capture or kill their members, and, ultimately, eradicate the group. Unlike the military, the police usually have a permanent presence in cities, towns, and villages; a better understanding of local groups and the threat environment in these areas; and better intelligence. The mission of the police and other security forces should be to eliminate the insurgent organization—the command structure, terrorists, logistics support, and financial and political support—from the midst of the population. As Bruce Hoffman argued, a critical step in countering terrorist groups is for law-enforcement officials to

> develop strong confidence-building ties with the communities from which terrorists are most likely to come or hide in, and mount communications campaigns to eradicate support from these communities. The most effective and useful intelligence comes from places where terrorists conceal themselves and seek to establish and hide their infrastructure.[49]

Police and intelligence services are best placed to implement these activities.

A police approach can include a range of steps. The first is intelligence collection and analysis. In *Small Wars*, British Colonel C. E. Callwell wrote that security forces often work somewhat "in the dark" against terrorist organizations and that "what is known technically as 'intelligence' is defective, and unavoidably so."[50] Intelligence is the principal source of information on terrorists. The police and intelligence agencies have a variety of ways to identify terrorists, including signal

---

[49] Hoffman (2006, p. 169).

[50] C. E. Callwell, *Small Wars: Their Principles and Practices*, 3rd ed., Lincoln, Neb.: University of Nebraska Press, 1996, p. 43.

intelligence (such as monitoring phone calls) and human intelligence (such as using informants to penetrate terrorist cells). Human intelligence can provide some of the most useful actionable intelligence. But it requires painstaking work and patience in recruiting informants who are already in terrorist organizations or placing informants in terrorist organizations. Intelligence collection and analysis can also be reinforced by countering the ideology and messages of terrorist groups through what is often referred to as *information operations*: the use of a variety of strategies and tools to counter, influence, or disrupt the message and operation of terrorist groups.[51]

Second is the arrest of key leaders and their support network. In democratic countries, this involves capturing key members and presenting the evidence in court. Terrorism involves the commission of violent crimes, such as murder and assault. Consequently, the investigation, trial, and punishment of perpetrators should be a matter for the wider criminal-justice system—including the police.[52] The barriers can be significant. Finding sufficient evidence that can be presented in court but that does not reveal sensitive information about sources and methods can be challenging. This is especially true if a terrorist has not yet perpetrated an attack. In many cases, it may be easier and more effective to arrest and punish terrorists for other offenses, such as drug trafficking, that may be ancillary to their terrorist activity. In nondemocratic societies, the policing approach is often drastically different, because laws and norms of behavior may be different.[53]

The third step is the development and passage of antiterrorism legislation. This can involve criminalizing activities that are necessary for terrorist groups to function, such as raising money or recruiting members. It can also involve passing laws that make it easier for intelli-

---

[51] On the use of *information operations* in a military context, see, for example, U.S. Joint Chiefs of Staff, *Information Operations*, Washington, D.C., February 13, 2006.

[52] See, for example, Clutterbuck (2004, pp. 142–144).

[53] On the dilemmas of U.S. assistance to police in nondemocratic countries, see Seth G. Jones, Olga Oliker, Peter Chalk, C. Christine Fair, Rollie Lal, and James Dobbins, *Securing Tyrants or Fostering Reform? U.S. Internal Security Assistance to Repressive and Transitioning Regimes*, Santa Monica, Calif.: RAND Corporation, MG-550-OSI, 2006.

gence and police services to conduct searches, engage in electronic surveillance, interrogate suspects, and monitor groups that pose a terrorist threat. It can include efforts to protect witnesses, juries, and judges from threats and intimidation. In democratic states, this inevitably leads to tension between civil liberties and security. Paul Wilkinson noted that the "primary objective of counter-terrorist strategy must be the protection and maintenance of liberal democracy and the rule of law."[54] The policing approach can also include providing foreign assistance to police and intelligence services abroad to improve their counterterrorism capacity.

In sum, a policing approach involves collecting intelligence on terrorists and terrorist groups (including their broader network), capturing key leaders, and passing legislation to prevent groups from operating and raising money.

---

[54] Wilkinson (1986, p. 125).

CHAPTER FOUR

Politics and the FMLN in El Salvador

In a solemn ceremony in Mexico City's Chapultepec Castle, representatives of El Salvador's government and the FMLN signed a peace settlement in January 1992. The agreement ended 12 years of civil conflict that left approximately 75,000 people dead and spelled the end of the FMLN as a terrorist organization.[1] The FMLN employed a variety of tactics, such as kidnappings, arson, and bombings, to coerce the Salvadoran government into making significant political, social, and economic changes. They targeted government officials and members of El Salvador's oligarchy. During an offensive operation in November 1989, for example, FMLN fighters occupied several upper-class neighborhoods in the capital city of San Salvador. As one FMLN member remarked, "By taking this part, we bring the war to the rich." The FMLN's clandestine radio station repeated this message in its broadcast to the city's residents: "We're already sitting in the oligarchy's best mansions."[2]

How did the FMLN end as a terrorist organization? This chapter argues that the possibility of a political settlement with terrorist groups is linked to a key variable: the breadth of terrorist goals. Most

---

[1] On the number of people killed during El Salvador's civil war, see David H. McCormick, "From Peacekeeping to Peacebuilding: Restructuring Military and Police Institutions in El Salvador," in Michael W. Doyle, Ian Johnstone, and Robert C. Orr, eds., *Keeping the Peace: Multidimensional UN Operations in Cambodia and El Salvador*, New York: Cambridge University Press, 1997, pp. 282–311, p. 282.

[2] Douglas Grant Mine, "Guerrillas Attack Affluent Neighborhoods," Associated Press, November 29, 1989.

terrorist groups that end because of politics seek narrow policy goals. The FMLN's goals, which were tied to political, social, and economic reforms in El Salvador, were largely policy oriented. The FMLN pushed for the transition to a democratic political regime and the termination of the Salvadoran government's repressive security apparatus, as well as such changes as land reform. The Salvadoran government was ultimately willing to negotiate from this baseline. The broader lesson for politics is straightforward: The narrower the goals of terrorist organizations, the more likely the government and terrorist group may be to agree on a settlement. This bargaining space does not exist with many terrorist groups. As the concluding chapter notes, for instance, al Qa'ida's broad goals of establishing a caliphate across the Middle East offer no bargaining room with western governments.

This chapter is organized into four sections. The first outlines the evolution of the FMLN and its primary policy goals. The second section examines the negotiation process between the FMLN and the government. The third examines the implementation of the agreement and the end of the FMLN. And the fourth section outlines the logic of when and why terrorist groups end because of politics.

FMLN Evolution and Goals

The FMLN's primary goals were policy-oriented. They included coercing the Salvadoran government to transition to a democratic political system and ending its repressive internal security apparatus, as well as instituting key social and economic changes, such as land reform. These goals had deep roots in El Salvador's history.

FMLN's origins go back more than a century, to at least the expansion of coffee cultivation in the late 1880s. Central to the evolution of El Salvador's political economy was a class structure based on the coercion of agrarian labor. State political elites enforced repressive labor conditions and highly concentrated property rights on behalf of a small economic elite. Despite a degree of economic diversification and modernization after World War II and despite several attempts at reform by regime moderates, coalitions of economic elites and military hardliners defended labor-repressive institutions and practices until FMLN's emergence. Salvadoran history can thus be characterized by elite resistance to change. It resulted in an economy based on export agriculture, elite reliance on the political control of labor, and elite opposition to political reform.[3] By the 1970s, the quiescence of those who were economically and politically excluded gave way to protest and political mobilization. Urban and rural youth found little opportunity for upward mobility, and social actors began to contest economic and political exclusion. For example, some elements of the Catholic Church began to encourage *campesinos* (farm laborers) to reflect on the Bible's implications for contemporary issues of social justice.[4] Such organizations as the Christian Federation of Salvadoran Peasants demanded land and better working conditions for *campesinos.* And Christian Democrats extended their party organization to the countryside from its origins among urban professionals.

As political mobilization increased and developed a national following, the Salvadoran government responded with a wave of repression. On February 28, 1977, for example, security forces opened fire on a midnight vigil following a political rally, leaving dozens dead.[5] Later in the year, repression deepened after the government suspended constitutional protections. Violence against progressive Catholics escalated as well. Between 1977 and 1979, six priests were killed.[6] In addition to controlling the army, El Salvador's military controlled the national guard, treasury police, national police, national intelligence directorate, and paramilitary civil-defense forces.[7] These forces functioned as

[3] Wood (2000, pp. 25–51).

[4] Carlos Rafael Cabarrús, *Génesis de una Revolución: Análisis del Surgimiento y Desarrollo de la Organización Campesina en El Salvador*, Mexico: Centro de Investigaciones y Estudios Superiores en Antropología Social, 1983; Rodolfo Cardenal, *Historia de una Esperanza: Vida de Rutilio Grande*, San Salvador, El Salvador: UCA Editores, 1985.

[5] William Deane Stanley, *The Protection Racket State: Elite Politics, Military Extortion, and Civil War in El Salvador*, Philadelphia, Pa.: Temple University Press, 1996, pp. 109–110.

[6] Anna Lisa Peterson, *Martyrdom and the Politics of Religion: Progressive Catholicism in El Salvador's Civil War*, Albany, N.Y.: State University of New York Press, 1997, pp. 63–65.

[7] El Salvador's security forces included 40,000 Army, 1,200 Navy, 2,400 Air Force, 4,000 national guard, 6,000 national police, 2,000 treasury police, and 24,000 civil-defense forces

political police and ruthlessly suppressed dissent throughout the country. Targets included labor and peasant organizations, church officials, religious workers, political opponents, the media, and human-rights monitors.[8] In his homily on March 23, 1980, Oscar Romero, the archbishop of San Salvador, delivered a striking rebuke to the Salvadoran government. The government viewed his increasingly vocal opposition as a threat to its power. The day after he delivered his speech, the archbishop was murdered. In the speech, Archbishop Romero had argued,

> No government can be efficient without the support of the people, least of all when it tries to enforce them through blood and pain. . . . I would like to make a special call to the men in the Army. . . . Brothers, you are from our people, killing your own peasant brothers, and before being given an order to kill by a man, the law of God must prevail, which says: Do not kill. . . . No soldier is obliged to obey an order against the law of God. . . . An immoral law, nobody has to accomplish it. . . . It is time to recover your consciousness and to obey your consciousness before a sinful order. . . . In the name of God, and in the name of this suffering people whose laments rise up to heaven, every day more tumultuous, I entreat and beg you, I order you in the name of God: Stop the repression![9]

The brutal response by regime elites to increasing political mobilization by peasants, workers, and students had several consequences. First, despite enduring differences in ideology and strategy, El Salvador's armed revolutionary groups unified their political representation in 1980 to form the FMLN.[10] This involved the integration of the Popu-

---

(International Institute for Strategic Studies, *The Military Balance, 1991–1992*, London: Brassey's, 1991, p. 198).

[8] Americas Watch Committee, *El Salvador's Decade of Terror: Human Rights Since the Assassination of Archbishop Romero*, New Haven, Conn.: Yale University Press, 1991, pp. 17–63.

[9] Jorge Caceres Prendes, "Revolutionary Struggle and Church Commitment: The Case of El Salvador," *Social Compass*, Vol. 30, Nos. 2–3, 1983, pp. 261–298, p. 293.

[10] Hugh Byrne, *El Salvador's Civil War: A Study of Revolution*, Boulder, Colo.: Lynne Rienner Publishers, 1996.

lar Liberation Forces, Popular Revolutionary Army, Communist Party of the Armed Forces of Liberation, National Resistance, and Workers Revolutionary Party. The FMLN began a prolonged campaign of terror and guerrilla warfare against the Salvadoran government and its civilian supporters.[11] Second, repression caused many people active in peasant or student organizations to support these previously inconsequential organizations. In interviews, several FMLN members stated that they had joined the guerrillas out of outrage at security forces' actions against family members or neighbors. Some joined the FMLN in response to the killing of priests, particularly the assassination of Archbishop Romero.[12] As one Salvadoran noted, "Before the war, we were despised by the rich. We were seen as animals, working all day and without even enough to put the kids in school. This is the origin of the war: there was no alternative. The only alternative was the madness of desperation."[13]

The FMLN argued that the "transformation of our society, [which] to date [has been] subjected to injustice, [as well as] the pillaging and selling out of our country, is today a possible reality close at hand. Only through this transformation will our people prevail and ensure the democratic freedoms and rights that they have been denied."[14] The success of the Sandinistas in overthrowing the Nicaraguan government in 1979 provided inspiration to FMLN members. The FMLN's goals can be divided into three components:

---

[11] The FMLN was named for the rebel leader Farabundo Martí, who led workers and peasants in an uprising to transform Salvadoran society after the eruption of the volcano Izalco in 1932. In response, the military regime led by General Maximiliano Hernández Martínez, who had seized power in a 1931 coup, launched a brutal counterinsurgency campaign that that killed 30,000 suspected guerrillas and Martí supporters.

[12] Carlos María Vilas, *Between Earthquakes and Volcanoes: Market, State, and the Revolutions in Central America*, New York: Monthly Review Press, 1995; Cynthia McClintock, *Revolutionary Movements in Latin America: El Salvador's FMLN and Peru's Shining Path*, Washington, D.C.: U.S. Institute of Peace Press, 1998.

[13] Elisabeth Jean Wood, *Forging Democracy from Below: Insurgent Transitions in South Africa and El Salvador*, Cambridge and New York: Cambridge University Press, 2000, p. 48.

[14] Torres-Rivas (1997, p. 220).

- *democratic transition:* transition from an authoritarian to a democratic political system
- *end of repression:* terminate government repression by the army and internal security forces and hold the government accountable for its human-rights violations
- *land reform:* establish land reform and improve conditions for *campesinos.*

To accomplish these objectives, the FMLN adopted a campaign of terror and insurgency against the government and elites. It employed a variety of tactics, such as kidnappings, arson, and bombings, to coerce the Salvadoran government into making significant political, social, and economic changes.[15] By the time of the 1992 peace agreement, the FMLN included more than 12,000 combatants, operated in all 14 provinces of the country, and controlled one-third of the country's territory.[16] FMLN guerrillas were capable of conducting major combat operations throughout El Salvador and, in 1989, had captured sections of San Salvador. They enjoyed strong popular support in certain areas of the country; a de-facto sanctuary in border areas disputed by El Salvador and Honduras; and a strong network of international financial, logistical, and political support.[17]

Negotiating an End to Terrorism

How did the FMLN end as a terrorist organization? The primary reason was that the FMLN's goals, which were tied to political and economic reforms, were narrow enough to allow for negotiating room.

---

[15] Call (2002).

[16] Charles T. Call, "Democratisation, War and State-Building: Constructing the Rule of Law in El Salvador," *Journal of Latin American Studies*, Vol. 35, No. 4, November 2003, p. 831; Call (2002, p. 386).

[17] Americas Watch Committee (1991, pp. 64–70); United Nations Department of Public Information, *The United Nations and El Salvador, 1990–1995*, New York, 1995a, p. 8.

Window of Opportunity

In leading up to this grand bargain, however, two structural conditions created a window of opportunity for negotiations. Neither of these conditions made a settlement inevitable, but they did create an important opportunity. First, the end of the Cold War created an opportunity for peace negotiations. As elsewhere in Central America, the war in El Salvador had evolved into a proxy conflict between the United States and Soviet Union. The Reagan administration viewed El Salvador as a place to draw the line against communist aggression and provided more than $6 billion in economic and military assistance to El Salvador's government over the course of the war.[18] But the end of the Cold War had a significant impact on the impetus for negotiations. The Soviet Union's withdrawal of support for Marxist movements in Latin America eliminated an important source of supply of arms and logistical support to the FMLN.[19] And the new administration with the election of George H. W. Bush as president put significant pressure on El Salvador's government to negotiate a peace settlement.[20] It threatened to withdraw aid if there was no settlement and offered financial assistance if a settle-

---

[18] Benjamin Schwarz, *American Counterinsurgency Doctrine and El Salvador: The Frustrations of Reform and the Illusions of Nation Building*, Santa Monica, Calif.: RAND Corporation, R-4042-USDP, 1991, p. 2. Also see Michael Childress, *The Effectiveness of U.S. Training Efforts in Internal Defense and Development: The Cases of El Salvador and Honduras*, Santa Monica, Calif.: RAND Corporation, MR-250-USDP, 1995; U.S. Department of State, *Communist Interference in El Salvador: Documents Demonstrating Communist Support of the Salvadoran Insurgency*, Washington, D.C., 1981.

[19] The FMLN also received economic and military assistance from Cuba and the Sandinistas in Nicaragua (Paulo S. Wrobel and Guilherme Theophilo Gaspar de Oliverra, *Managing Arms in Peace Processes: Nicaragua and El Salvador*, New York: United Nations, 1997, p. 124). Also see Mark Levine, "Peacemaking in El Salvador," in Michael W. Doyle, Ian Johnstone, and Robert C. Orr, eds., *Keeping the Peace: Multidimensional UN Operations in Cambodia and El Salvador*, New York: Cambridge University Press, 1997, pp. 227–255, pp. 230–231; and Gerardo L. Munck and Dexter Boniface, "Political Processes and Identity Formation in El Salvador: From Armed Left to Democratic Left," in Ronaldo Munck and Purnaka L. De Silva, eds., *Postmodern Insurgencies: Political Violence, Identity Formation, and Peacemaking in Comparative Perspective*, New York: St. Martin's Press, 2000, pp. 38–53.

[20] Terry Lynn Karl, "El Salvador at the Crossroads: Negotiations or Total War," *World Policy Journal*, Vol. 6, No. 2, April 1989, pp. 321–355; Wood (2000, p. 81).

ment was reached.[21] In addition, the leaders of four major countries—Spain, Colombia, Mexico, and Venezuela—approached the United Nations for assistance in reaching a negotiated settlement to the war.

Second, the Salvadoran civil war was at a stalemate. Neither side had achieved a military victory under the prevailing political conditions, and it was unclear whether either side *could* achieve a victory in the foreseeable future. After a few initial meetings between the FMLN and government representatives in mid-1989, the process ground to a halt, and, in a context of increasing political violence, the FMLN launched a major offensive in November 1989. Although the group did not hold any city for more than a few weeks, it brought the war home to the wealthy neighborhoods of San Salvador, underscoring the inability of the Salvadoran military to contain the war. The assassination of six Jesuits and their two female employees by the Atlacatl Battalion, an elite unit of the Salvadoran armed forces, during the rebel offensive made further U.S. congressional support untenable in the absence of peace negotiations.[22]

Negotiating a Settlement

Negotiations were possible because the FMLN's goals were sufficiently narrow. Each step made some progress on the key elements of the negotiation—democratic transition, end of repression, and land reform. The bargaining process took place in five major stages.

The first was the Geneva Agreement of April 1990, which became the cornerstone on which the negotiating process developed. The FMLN and the Salvadoran government agreed to begin a peace process whose outcome would be marked by the end of arm conflict, promotion of democracy, guarantee of human rights, and integration of FMLN combatants into civilian life. The agreement also created

---

[21] Torres-Rivas (1997).

[22] Teresa Whitfield, *Paying the Price: Ignacio Ellacuría and the Murdered Jesuits of El Salvador*, Philadelphia, Pa.: Temple University Press, 1994; Mark O. Hatfield, James Leach, and George Miller, *Bankrolling Failure: United States Policy in El Salvador and the Urgent Need for Reform: A Report to the Arms Control and Foreign Policy Caucus*, Washington, D.C., 1987.

a negotiating framework under the auspices of the United Nations Secretary-General.[23]

The second stage led to an agreement in San Jose, Costa Rica, in July 1990. The military was a major topic of discussion. The FMLN viewed the military through the prism of 60 years of military domination of the country's politics, ten years of civil war, and innumerable human-rights abuses. Consequently, FMLN negotiators pushed for the abolishment of the armed forces. But the two sides were unable to reach an agreement on the army. Instead, they settled on a human-rights agreement.[24] The San Jose agreement provided for the establishment of a United Nations verification mission to monitor nationwide respect for and the guarantee of human rights and fundamental freedoms in El Salvador. The focus of the agreement was on establishing accountability for the Salvadoran government's gross human-rights violations. Both parties agreed to the establishment of a United Nations Observer Mission in El Salvador (ONUSAL) to verify compliance with the accords' human-rights provisions. According to the agreement, ONUSAL was to take up its duties as of the cessation of the armed conflict. Shortly after signing the agreement, however, the two parties requested that the UN mission be set up even *before* a cease-fire, leading the United Nations Secretary-General to send a preliminary mission in March 1991.[25]

The third stage occurred in April 1991, when the two parties agreed on a set of changes to the 1983 constitution in the April 27 Mexico Agreements. The most important changes concerned the institutional mandate of the armed forces. The military's mission was explicitly limited to external defense, and policing was to be under civilian control. The agreement led to the "creation of a National Civil

---

[23] Ricardo G. Castaneda, "Letter Dated 91/10/08 from the Permanent Representative of El Salvador to the United Nations Addressed to the Secretary-General," New York: United Nations, A/46/551-S/23128, October 9, 1991b.

[24] Levine (1997, pp. 233–234).

[25] Alvaro de Soto, representative of the secretary-general of the United Nations, "Note Verbale Dated 90/08/14 from the Charge D'Affaires A.I. of the Permanent Mission of El Salvador to the United Nations Addressed to the Secretary-General," New York: United Nations, A/44/971-S-21541, August 16, 1990.

Police for the maintenance of peace, tranquility, order and public safety in both urban and rural areas, under the control of civilian authorities." It noted that "the National Civil Police and the armed forces shall be independent and shall be placed under the authority of different ministries."[26] The constitutional agreement also provided for a truth commission under United Nations auspices to investigate past human-rights violations. Despite some opposition by hard-line members of the governing party (the National Republican Alliance Party, or Alianza Republicana Nacionalista [ARENA]), the agreements were subsequently ratified by the ARENA-controlled national assembly, an indication of Salvadoran president Alfredo Cristiani's ability to deliver party compliance on at least some of the reforms, despite opposition from party hardliners. But there were numerous issues on which the two sides could not agree, such as land reform and the logistics of a cease-fire.

The fourth stage was the New York Agreement, which was signed on September 25, 1991. In September 1991, both the United States and the Soviet Union had expressed their support for the negotiations and contributed pressure to reach an agreement. The FMLN gave up its long-standing demand to merge with or disband the army and agreed that no new issues would be added to the cease-fire talks. The Salvadoran government agreed to protect the right of rebel families to hold onto land they had occupied during the war. Some FMLN members would join the new civilian police force, the National Civilian Police; the armed forces would be reduced in size and the treasury police and national guard eliminated; and the military officer corps would be purged by a newly created ad hoc commission.[27] It eventually issued a report on September 22, 1992, following a review of 232 of the

---

[26] Soto (1990).

[27] President Cristiani initially refused to remove all of the named officers. After pressure from the UN Secretary-General, the United States, and several neighboring countries, Cristiani ultimately acceded to the recommendation, although he transferred some officers to embassies abroad and allowed others to retire (Boutros Boutros-Ghali, "Letter Dated 93/01/07 from the Secretary-General Addressed to the President of the Security Council," New York: United Nations, S/25078, January 9, 1993b; Ian Johnstone, "Rights and Reconciliation in El Salvador," in Michael W. Doyle, Ian Johnstone, and Robert C. Orr, eds.,

most senior military officers. The report recommended the discharge of the entire senior military establishment, including officers who had played an integral role in the peace process. The New York Agreement also included the first outline of land redistribution and economic and social reforms. State or private land holdings that exceeded 245 hectares, as well as land that was not distributed for ecological preservation, would be subject to redistribution.[28]

The fifth and final stage culminated in the January 1992 Chapultepec Peace Accords. It was a political compromise in which the FMLN agreed to a democratic political regime and a capitalist economy with limited socioeconomic reform, and the Salvadoran government agreed to the FMLN's political participation along with some socioeconomic reform. The agreement enshrined a democratic bargain: The two sides agreed to resolve their future differences through democratic political processes. Consequently, the agreement's principal achievement was an agenda of reforms that would institutionalize the new—and democratic—rules of the political game. The main provisions, many of which were carried over from the New York agreement, called for reform of the armed forces, accountability for past human-rights violations, the founding of a civilian police force, and restriction on the arbitrary exercise of state power. The FMLN participated in the general elections of March 1994, making a respectable showing at the presidential and legislative levels. The military was restructured, with more than 100 officers forced into retirement by the work of the ad hoc commission and reduced in size. The long-standing, close relationship between landlords and local military and police authorities was ended, and a new civilian police force was deployed throughout the country. The programs for former combatants proved more effective as economic assistance in the short run than as foundations for adequate

---

Keeping the Peace: Multidimensional UN Operations in Cambodia and El Salvador, New York: Cambridge University Press, 1997, pp. 312–341, pp. 316–318).

[28] Ricardo G. Castaneda, ambassador and permanent representative, "Letter Dated 91/09/26 from the Permanent Representative of El Salvador to the United Nations Addressed to the Secretary-General," New York: United Nations, A/46/502-S/23082, September 26, 1991a.

livelihoods in the long run. But combatants returned to civilian life, and political violence declined.[29]

Implementing the Agreement

Over the subsequent several years, the FMLN demobilized its forces and transitioned from a terrorist and insurgent organization to a viable political party. It remained committed to nonviolence because its key goals were met over the long run: democratic transition, end of government repression, and some land reform. This was not inevitable. As Barbara Walter has pointed out, peace settlements have often broken down between warring parties—including between terrorist organizations and the government.[30] A failure to implement the Chapultepec agreement would likely have triggered a return to terrorism.

Democratic Transition

The United Nations monitored and supported the conduct of four simultaneous elections in March 1994: for president, parliament, municipal councils, and the Parlamento Centroamericano. El Salvador's Tribunal Supremo Electoral managed all the elections.

ONUSAL's electoral division assisted in voter registration, monitored the election campaign, and provided assistance in drawing up the voter rolls. During the voter-registration period, ONUSAL offered technical and logistical support to the Tribunal Supremo Electoral, which possessed outdated computer equipment and faced transportation and communication problems.[31] Monitoring teams made more than 2,350 visits to towns throughout the country to assist in voter

---

[29] Graciana Del Castillo, "The Arms-for-Land Deal in El Salvador," in Michael W. Doyle, Ian Johnstone, and Robert C. Orr, eds., *Keeping the Peace: Multidimensional UN Operations in Cambodia and El Salvador*, New York: Cambridge University Press, 1997, pp. 342–366, pp. 362–363; Wood (2000, p. 107).

[30] Barbara F. Walter, *Committing to Peace: The Successful Settlement of Civil Wars*, Princeton, N.J.: Princeton University Press, 2002.

[31] Boutros Boutros-Ghali, "Report of rhe Secretary-General on the United Nations Observer Mission in El Salvador," New York: United Nations, S/1994/179, February 6, 1994a, p. 3.

registration.[32] Observer teams attended more than 800 political events and monitored political advertising through the mass media. On election day, ONUSAL deployed almost 900 international observers to observe conduct at the polls and the counting of the ballots; each of the 355 policing centers in El Salvador was monitored. Following the UN recommendation, polling stations marked voters with indelible ink to prevent multiple voting by the same individual.[33] A team of 40 specialized observers was also deployed to the Tribunal Supremo Electoral to monitor the official count.[34] There were some voting irregularities and organizational problems on election day. But the elections were ultimately successful. There were no serious security problems and no ballot rigging. ARENA received 49 percent of the vote and 39 seats in the legislative assembly, the FMLN coalition received 25 percent and 22 seats, and the Christian Democratic Party 5 percent and 18 seats. No candidate in the presidential election obtained an absolute majority. A second round of voting was held on April 24, 1994, between the two candidates with the highest number of votes. Armando Calderón Sol, the ARENA candidate, won with 68 percent.

Over the ensuing decade, El Salvador regularly held free and fair democratic elections. Newspapers, books, magazines, films, and plays were not censored. Academic freedom was respected. The FMLN reinvented itself as a political party and became an increasingly powerful political force. It increased its seats in the national assembly in 1997; became the largest party in the assembly in 2000; and retained that position in 2003, with 35 percent of the popular vote.[35]

---

[32] Boutros Boutros-Ghali, "Report of the Secretary-General on the United Nations Observer Mission in El Salvador," New York: United Nations, S/1994/304, March 16, 1994b, p. 2; UN/DPI (1995a, p. 438).

[33] Boutros-Ghali (1994a, p. 6; 1994b, p. 4).

[34] Boutros Boutros-Ghali, "Report of the Secretary-General on the United Nations Observer Mission in El Salvador," New York: United Nations, S/1994/375, March 31, 1994c, p. 3.

[35] Jenny Pearce, "From Civil War to 'Civil Society': Has the End of the Cold War Brought Peace to Central America?" *International Affairs*, Vol. 74, No. 3, July 1998, pp. 587–615, p. 605.

End of Repression

ONUSAL's military division successfully oversaw and verified the dissolution of the FMLN's military structure, destruction of its weapons and equipment, and its transition from a combatant force to a political party. ONUSAL encountered several problems along the way. It verified that the FMLN had destroyed or handed over all weapons in December 1992, when it formally announced the end of armed conflict.[36] But the accidental explosion of an undisclosed arms cache in May 1993 and the discovery of large quantities of weapons indicated that the FMLN had not handed in all its weapons.[37] Over the succeeding months, the FMLN informed ONUSAL of another 114 arms caches in El Salvador, Nicaragua, and Honduras. These contained ammunition, rockets, grenades, and surface-to-air missiles. In total, the FMLN destroyed 10,230 weapons, 140 rockets, 9,228 grenades, 5,107 kilograms of explosives, 74 surface-to-air missiles, and more than 4 million rounds of ammunition.[38]

Efforts to reform the armed forces were also largely successful, though there were notable challenges. Under ONUSAL supervision, the government demobilized the civil-defense patrols and reduced the size of the army from 40,000 to 28,000 soldiers. U.S. military advisers trained and assisted with restructuring. They helped develop a new training and doctrine command and provided technical advice on the reorganization of El Salvador's military college.[39] Instead of abolishing the national guard and treasury police, as stipulated in the Chapultepec agreement, the government incorporated those organizations structurally intact into the army and renamed them the National Border Guard and Military Police, respectively. They retained largely the same missions.

---

[36] Boutros Boutros-Ghali, "Report of the Secretary-General on the United Nations Observer Mission in El Salvador (Onusal)," New York: United Nations, S/25006, December 23, 1992, pp. 282–285.

[37] Boutros-Ghali (1992; 1995, p. 2).

[38] Wrobel and Gaspar de Oliverra (1997, pp. 137–138).

[39] McCormick (1997, p. 297).

Land Reform

ONUSAL was partially successful in reintegrating former combatants into Salvadoran society. The Chapultepec agreement provided for the transfer of land to former FMLN combatants, members of the armed forces, and squatters who had occupied land in the conflict areas during the war.[40] A maximum of 7,500 former FMLN combatants, 15,000 former military soldiers, and 25,000 landholders in the zone of conflict were to be reintegrated.[41] They were given credit to purchase land, agricultural training, basic household goods, agricultural tools, housing, and technical assistance. The available land would come from territory that landowners abandoned or were forced off of during the civil war. The National Commission for the Consolidation of Peace (Comisión Nacional para la Consolidación de la Paz, or COPAZ) was responsible for administering the land-transfer program. But the agreement's vagueness on several issues created numerous problems. It said nothing, for example, about the size of the plots to which the beneficiaries were entitled, the amount of government credit available to beneficiaries, and the practical arrangements under which the land was to be transferred.[42] COPAZ was also ineffective at mediating land disputes. By the end of 1994, land titles had been issued to only 40 percent of potential beneficiaries. Setbacks continued over the subsequent several years for a number of reasons: coordination problems among Salvadoran government agencies, payment delays, legal problems, and some owners' refusal to sell their land.[43]

---

[40] UN/DPI (1995a, pp. 206–209).

[41] UN/DPI (1995a, p. 29).

[42] Timothy A. Wilkins, "The El Salvador Peace Accords: Using International and Domestic Law Norms to Build Peace," in Michael W. Doyle, Ian Johnstone, and Robert C. Orr, eds., *Keeping the Peace: Multidimensional UN Operations in Cambodia and El Salvador*, New York: Cambridge University Press, 1997, pp. 275–277.

[43] Del Castillo (1997, pp. 342–365).

The Logic of a Political Solution

The FMLN experienced a fundamental transition from a terrorist organization to a political party in the 1990s. By the late 1980s, structural conditions created a window of opportunity for a settlement. The Soviet Union's withdrawal of support for Marxist movements in Latin America was a major blow to the FMLN, and the civil war had reached a stalemate. In this context, the FMLN's narrow goals made a negotiated settlement possible. Both sides had to make significant compromises. The FMLN did not get as much military or land reform as it had initially hoped. And the Salvadoran government agreed to more military and police reform, greater transparency of its human-rights abuses, and a more democratic political system than it had initially wanted. But the outcome was successful: the end of the FMLN as a terrorist organization.

What does the FMLN experience suggest about the end of terrorist groups more broadly? First, structural developments can create a window of opportunity for a settlement, such as pressure from outside powers or a stalemate between the government and the terrorist organization. As one study on peace settlement concluded, combatants almost always returned to violence "unless a third party stepped in to enforce or verify a post-treaty transition. If a third party assisted with implementation, negotiations almost always succeeded, regardless of the initial goals, ideology, or ethnicity of the participants. If a third party did not, these talks almost always failed."[44] The United States, regional powers, or even international organizations (such as the United Nations) can play an important role as a third party in helping enforce agreements. For example, the United States, Ireland, and outside bodies, such as the Independent International Commission on Decommissioning, played fundamental roles in the end of the Provisional IRA as a terrorist organization. Second, when terrorist groups have broad goals (such as regime change, empire, or the transformation of a country's social order) the likelihood of a settlement with the government is low. This means that there is little likelihood of a settle-

---

[44] Walter (2002, p. 3).

ment with such groups as al Qa'ida that advocate the establishment of a transnational caliphate across the Middle East. It is not possible for any single government to accede to this demand.

As noted in Chapter Two, the possibility of a political solution is linked to a key variable: terrorist goals. Most terrorist groups that end because of politics seek narrow policy goals, such as policy and territorial change. There are two distinct logics for why terrorist groups end because of politics.

First, the narrower the goals of terrorist organizations, the more likely the government and the terrorist group are able to find a mutually agreeable settlement. The reason is straightforward: The government may have less to lose when the terrorist group has narrow goals. Agreeing to change a policy is easier to accomplish—and may be easier to sell to a domestic population—than agreeing to overturn an entire social order. When a terrorist group's goals are minimal, there may be a middle ground in which to negotiate a compromise settlement. As Figure 4.1 highlights, terrorist goals can range from narrow ones (such as a policy change) to broader ones (such as changing a country's social order). The further right on the x-axis, the broader the goals and the lower the state's willingness to reach a settlement. Consequently, the possibility of a settlement is greater the further left on the x-axis. As will be discussed in more detail in the next section, the FLMN's goals were largely policy oriented. A number of factors can influence bargaining positions in the middle of negotiations. One is a change in structural conditions, such as outside support or a transition from authoritarianism to democracy. For example, a decline in outside support from major powers can cut off a key source of funding and increase a group's willingness to settle. Another factor is a military stalemate. The inability of either side—the government or the terrorist group—to defeat the other over a period of years may also increase the likelihood of a negotiated settlement. The financial and blood costs of continuing to fight a protracted, unwinnable war may be too high.

Second, the broader the goals of terrorist organizations—such as regime change, empire, or social revolution—the more likely the government will be resistant to change, and the more likely that terrorism will be viewed as the only option. And the narrower the goals, the

Figure 4.1
FMLN Goals and Settlement Prospects

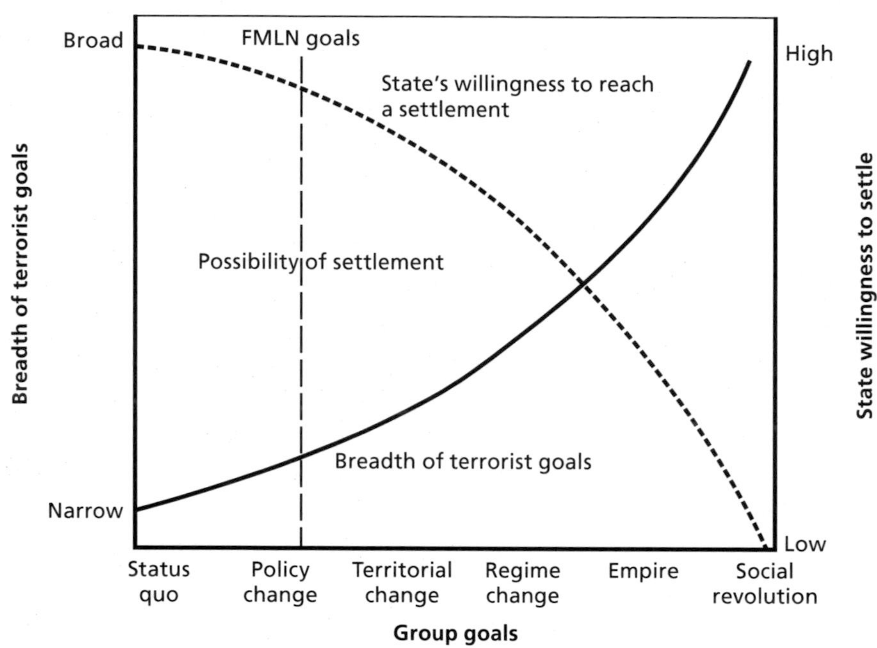

RAND *MG741-4.1*

more likely nonviolent tactics may be possible. This is particularly true if there are clear alternatives to terrorism, such as protests. As Martha Crenshaw noted, "opportunities for collective action, such as mass protest or revolution, may occur independently of government actions." Or "government repression may be a catalyst for the short of mass mobilization that makes terrorism unnecessary, or a transition to democracy may permit effective legal opposition."[45] Unlike the peace-settlement option, terrorist groups that pursue this path do not necessarily reach a formal agreement with the government. They calculate that pursuing their goals through nonviolent, civic means has greater benefits and lower costs. Indeed, most terrorist organizations are strategic calculators. They use violence to achieve specific political purposes, such as coercing a target government to change policy, mobilizing additional

---

[45] Crenshaw (1996, p. 266).

recruits and financial support, or achieving independence. They usually have a set of hierarchically ordered goals and choose strategies that best advance them. Consequently, they are influenced by cost-benefit calculations. Resorting to terrorism has benefits if groups can successfully achieve their goals. But it also has costs. Terrorists are constantly on the run because government security forces are trying to capture or kill them. Terrorism provokes repression that some organizations believe they cannot survive.

In sum, the likelihood that a terrorist group will end because of politics may be linked to the breadth of their goals. When their goals are narrow, either (1) the government is more likely to negotiate because it has less to lose or (2) nonviolent approaches may become plausible alternatives. FMLN's goals were sufficiently narrow that the government was willing to negotiate.

CHAPTER FIVE

Military Force and al Qa'ida in Iraq

U.S. operations in al Anbar province provide a useful illustration of when military forces can be appropriate against terrorist groups. While politics and policing may be more effective in most cases, military force can be critical when facing a terrorist group involved in an insurgency. Such groups are often well equipped, well organized, and well motivated, and police acting alone would be quickly overpowered.

This chapter argues that the U.S. military was effective in countering al Qa'ida in Iraq (AQI), notably in Iraq's al Anbar province. Though AQI had not been defeated at the time this book was published, its presence in al Anbar was a fraction of what it was at the height of its power in mid-2006. U.S. efforts from 2006 to 2008 illustrated how military forces might be used to counter terrorist groups involved in an insurgency. The United States was able to exploit AQI's growing unpopularity. It helped catalyze an organized opposition to AQI spearheaded by local tribes but supported in large numbers by disaffected fighters from other Sunni insurgent groups. AQI had essentially brought its troubles onto itself through an overly aggressive stance in al Anbar province. In a nutshell, what was a cooperating and networked insurgency prior to 2006 turned into an insurgency dominated by an increasingly confident AQI that was coming into power in the province. Using a variety of methods, AQI began to squeeze out other insurgent groups. Upset and under pressure, members of the latter insurgent groups switched sides against AQI and joined ranks with anti-AQI tribes. AQI was thus largely driven out of al Anbar.

In examining the al Anbar case, this chapter is organized into five sections. The first examines the networked nature of the Iraq insurgency. The second section explores the shifting tide against AQI. The third section outlines the fight against al Qa'ida, especially by local tribes. The fourth section examines the U.S. role in countering al Qa'ida, especially the role of U.S. military forces. And the final section offers conclusions on military force and the end of terrorist groups.

A Networked Insurgency

The Sunni insurgency in Iraq was, for many years, a complex and difficult-to-penetrate endeavor. What was remarkable about the Sunni insurgency (especially compared with other insurgencies) was the extent to which its components defied identification and categorization. Within a few months after the fall of Saddam Hussein, the insurgency had organized itself as a tacit alliance of Ba'athists, salafists, nationalists, sectarians, and criminals. The insurgency had no leadership, much less one that could be identified and targeted. By 2004 salafists, notably those associated with Abu Musab al-Zarqawi, had taken a prominent role in the fighting. This was especially true in Fallujah, a city in al Anbar province. Figure 5.1 highlights Al Anbar and neighboring provinces.

Several snapshots of the period suggest the difficulty of clearly deducing who was who in the insurgency. In a 2003 memo to Secretary Rumsfeld, Paul Bremer, the administrator of the Coalition Provisional Authority (CPA) in Iraq, noted that the threat to U.S. forces came from several sources. The first were elements of the former regime, including Ba'athists, Fedayeen Saddam, and intelligence agencies. They focused their attacks on three targets: coalition forces; infrastructure; and employees of the coalition, including Iraqis. "To date," Bremer wrote, "these elements do not appear to be subject to central command and control. But there are signs of coordination among them." The second threat was Iranian subversion: "Elements of the Tehran government are actively arming, training and directing militia in Iraq. To date, these armed forces have not been directly involved in attacks on the

Figure 5.1
Map of al Anbar Province and Vicinity

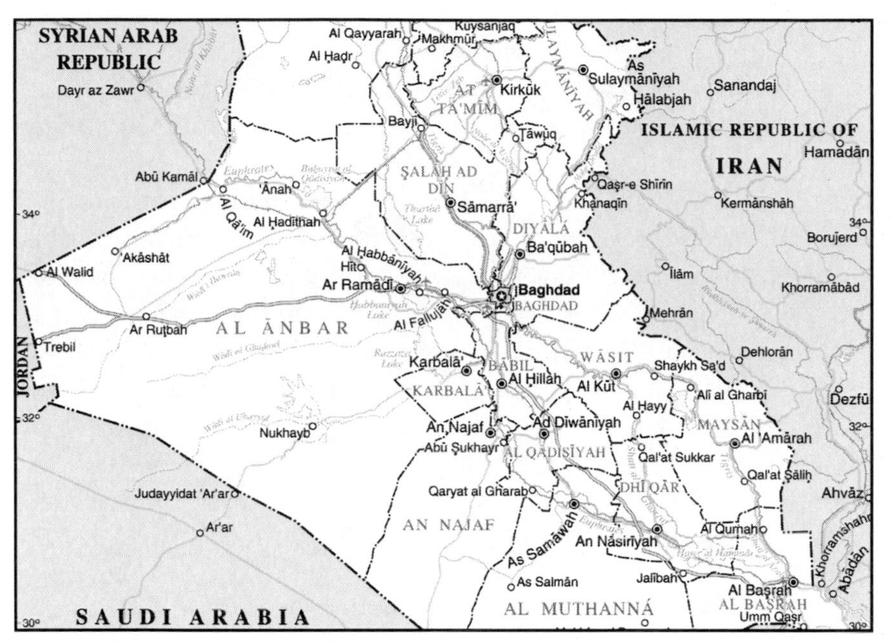

SOURCE: UN Cartographic Section, UN Department of Peacekeeping Operations (2004). Used with permission.
RAND *MG741-5.1*

Coalition. But they posed a longer term threat to law and order in Iraq." The third threat included terrorist groups, especially jihadists from Saudi Arabia, Syria, or Yemen. Bremer noted that there were clear indications that Ansar al-Islam, a terrorist group associated with al Qa'ida, was operating inside Iraq and actively surveying coalition targets.[1]

An early 2006 report by the International Crisis Group observed, "The insurgency is built around a loose and flexible network [and] feeds on deep-seated family, tribal and local loyalties with allegiance to cause rather than to specific individuals." Legitimacy, which was of

---

[1] Paul Bremer, director of reconstruction and humanitarian assistance, "Message for SecDef," email to Jaymie Durnan, special assistant to the deputy defense secretary, June 30, 2003.

considerable importance to the insurgents, was based almost entirely on negative perceptions shared by the Sunni population and the insurgents, "opposition to the occupation, anger at its specific practices, and the feeling . . . of being under siege."[2] The International Crisis Group report further observed that, within the vast heterogeneous insurgency, four groups stood out: AQI, Jamaat Ansar al-Sunnah, the Islamic Army in Iraq (IAI), and the Islamic Front for Iraqi Resistance: Salah-Al-Din Al-Ayyubi Brigades (Al-Jabhah al-Islamiyah al-Muqawamah al-Iraqiyah, or JAMI). Minor groups identified by the International Crisis Group included Jaish al-Rashidin, Victorious Sect, Jaish al-Mujahidin, the 1920 Revolution Brigades (the military arm of Islamic Resistance Movement in Iraq), and JeM.[3]

In addition, other assessments identified the primary groups as JeM, IAI, Iraqi National Islamic Resistance, the Mujahedin Army, and Jamaat Ansar al-Sunnah. The insurgency's leadership consisted of a dozen or so individuals who met occasionally to discuss organization and tactics. As one assessment concluded,

> while the jihadists get the most attention—because of their emphasis on mass-casualty attacks, and because they take credit for almost every major attack that occurs—the Iraqi 'armed national resistance' is probably responsible for most attacks on coalition forces and Iraqis associated with the government.[4]

Based on a different source methodology (relying heavily on press announcements), Ahmed S. Hashim identified several insurgent groups: nationalist tribal groups, nationalist religious groups, and salafist groups. They included Jamaat Ansar al-Sunnah, Victorious Sect, Muja-

---

[2] International Crisis Group, *In Their Own Words: Reading the Iraqi Insurgency*, Amman and Brussels, 2006, p. 25.

[3] The International Crisis Group translated it as Victorious Group's Army; we use *Victorious Sect* to refer to this group.

[4] Michael Eisenstadt and Jeffrey White, *Assessing Iraq's Sunni Arab Insurgency*, Washington, D.C.: The Washington Institute for Near East Policy, Policy Focus 50, December 2005.

hideen Battalions, Jihad Brigades, al Qa'ida Fallujah Branch, JeM, and IAI. Hashim also made copious mention of Zarqawi's group.[5]

One of the salient and well-reported transitions within the insurgency was the rapid decline in the influence of Saddam loyalists over its first two years. Simultaneously, most nationalist groups turned toward salafist interpretations of Islam, factors that gave an increasingly salafist tone to the entire insurgent movement.[6] Tensions between the groups occasionally surfaced over tactics, the legitimacy of violence against certain targets, and their attitude toward the political process. Nevertheless, as the International Crisis Group report argued,

> what is remarkable is that . . . violent friction between groups, far from precipitating the insurgency's implosion, has increased its coherence, at least in rhetoric. . . . Eager for legitimacy and fearful of debilitating internal conflict, the insurgency converged around an Islamic discourse, turning principally to salafi . . . religious scholars for moral and juridical validation of its jihad in general and of specific forms of conduct in particular.[7]

For example, the various insurgent groups disagreed over whether and how violently to oppose voting in the January 2005 elections. When the violence against voters backfired, six of the most active organizations issued a joint communiqué against targeting voters, and the critical seventh organization, AQI, fell in line. The number of attacks on election days in October and December 2005 dropped significantly compared to what took place in the January 2005 elections. Even as late as spring 2006, Sunni jihadists were observed to be essentially unified and held to a common doctrine.[8]

---

[5] Ahmed Hashim, *Insurgency and Counter-Insurgency in Iraq*, Ithaca, N.Y.: Cornell University Press, 2006, pp. 170–176.

[6] ICG (2006, p. 7). The few groups that professed attachment to the former regime quickly vanished.

[7] ICG (2006, p. 10).

[8] See, for example, Mathieu Guidère and Peter Harling, "Iraq's Resistance Evolves," *Le Monde Diplomatique*, May 2006.

The perception among observers as 2005 ended was that a network of like-minded jihadists was carrying out the insurgency in Iraq, with AQI as perhaps first among equals. In October 2004, Zarqawi had pledged loyalty to bin Laden, and bin Laden returned the favor by anointing Zarqawi his representative. The association with bin Laden improved Zarqawi's access to outside sources of income, though the boost to recruiting was more modest, since Zarqawi already had a good pipeline. Yet, such an endorsement from the outside fed the perception that Zarqawi led a group that did not necessarily have Iraq's interests in mind.

The Tide Turns

AQI's thrust for leadership of the Sunni insurgent movement was first made evident in the January 2006 formation of the Mujahideen Shura Council (MSC). The council included AQI as its de facto core, plus Victorious Sect and four lesser-known allied groups.[9] It attempted to put an Iraqi face on an insurgency that was initiated by non-Iraqis (notably Zarqawi, a Jordanian). In October 2006, AQI formed yet another front group, merging MSC with four brigades (Fursan al-Tawhid, Knights of Monotheism, Millah Ibrahim, and Religion of Ibrahim) as well as some tribal groups.[10] The result was the creation of the Islamic State of Iraq (ISI), which was the first step toward al Qa'ida's goal of establishing a caliphate in the region. Effective assimilation meant the end of some insurgent groups as independent entities. The reference to brigades suggests that their efforts to assimilate complete insurgent groups, such as IAI or the Mujahideen Army had failed, but they were able to peel

---

[9] Not every group listed in the MSC roster (or its successor, ISI) had actually agreed to pledge loyalty.

[10] Although the original October 13th announcement claimed Al-Fatihin as an alliance member, Al-Fatihin denied as much, leading MSC to counterclaim that the denial came from only one of Al-Fatihin's five brigades. Al-Fatihin continued to produce its own attack claims without reference to ISI.

off some members. The formation of ISI did not end AQI's efforts to assimilate more Sunni insurgents.[11]

ISI's creation had several putative motivations. It was an attempt to trigger splintering and encourage other Sunni insurgent groups to pledge alliance to ISI. Additionally, by claiming to be a state, ISI apparently sought to gain legitimacy. It was seen as attempting to take the military and political initiative from the other Sunni terrorist groups.[12] Less publicized, but probably more important, was AQI's attempt to unify Sunni Arabs into an Islamic state by force by cracking down on traditional smuggling and assassinating recalcitrant sheikhs.[13]

To accomplish its goals, ISI used brutal tactics. Sheikh Harith Zaher al-Dhari, son of the head of the al-Zouba' tribe, was assassinated, an act attributed to his organization's refusal to join ISI.[14] Citing "fighters from various groups," *al-Hayat* reported that some 30 commanders of Kataib Thawrat al-Ishrin and al-Jaysh al-Islami were assassinated by AQI or killed in battles over arms caches in the al Anbar district, which AQI had been attempting to capture in prior months. Elsewhere, it observed, "al-Qaeda has waged a war of liquidation with the primary targets being the leaders of [1920 Revolution Brigades] and IAI" because ISI failed to convince these groups to unite under their own banner.[15] In October, Sheikh abu Osama al-'Iraqi called on Osama bin Laden to denounce AQI for harming Sunnis and targeting jihad fighters from other Sunni factions, claiming that they had attacked the leaders of the faction known as the Ishrin Revolution Brigades. AQI

---

[11] For instance, in mid-December, the al Anbar branch of the Islamic Army in Iraq announced that it was joining ISI.

[12] The circular went on to note "several reports in recent months of fighting between Sunni nationalist groups and the jihadists . . . al Qaeda and its jihadist allies now face problems from fellow jihadists as well as Sunni nationalists."

[13] Sydney J. Freedberg Jr., "The New Iraqi Way of War," *National Journal*, Vol. 39, No. 23, June 9, 2007, pp. 36–43, p. 43.

[14] Middle East Media Research Institute, *Continued Clashes in Iraq Between Sunni Jihad Groups and al-Qaeda*, Washington, D.C., special dispatch 1542, April 13, 2007b.

[15] Lydia Khalil, "The Islamic State of Iraq Launches Plan of Nobility," *Terrorism Focus*, Vol. 4, No. 7, March 27, 2007, p. 4.

had also killed leaders from other factions, leaders on whose heads U.S. forces had offered rewards of hundreds of thousands of dollars.[16]

At the same time that AQI was trying to muscle other Sunni insurgent groups to come under its banner, it sought to tighten control over al Anbar itself. The stronger AQI became, the harder it pushed on the various institutions associated with acquiring resources and recruits. Where it could, it began to impose its own laws and enforced them through Islamic courts. These courts implemented a strict version of shari'a. They banned cigarettes, imposed strict dress codes for women, and prohibited female drivers.[17] Marine Colonel Peter Devlin, writing on al Anbar, observed that AQI was an "integral part of the social fabric . . . the dominant organization of influence in al-Anbar [with] an ability to control the day-to-day life of the average Sunni."[18] AQI was also able to insert its members into the smuggling trade, notably in oil refining and oil distribution.[19]

Fighting al Qa'ida

AQI's use of murder and intimidation to win allies reached beyond Sunni insurgent groups and extended to the various tribes of al Anbar. Many tribes were initially opposed to the U.S. intervention but wary of a theocratic ideology that left little room for tribal authority. Although

---

[16] D. Hazan, *Sunni Jihad Groups Rise Up Against Al-Qaeda in Iraq*, Washington, D.C.: Middle East Media Research Institute, Inquiry and Analysis Series 336, March 22, 2007.

[17] Middle East Media Research Institute, *The Islamic State of Iraq Issues Regulations for Women Drivers*, Washington, D.C., special dispatch 1514, March 23, 2007a.

[18] Michael R. Gordon, "G.I.'s Forge Sunni Tie in Bid to Squeeze Militants," *New York Times*, July 6, 2007, p. A1; Dafna Linzer and Thomas E. Ricks, "Anbar Picture Grows Clearer, and Bleaker," *Washington Post*, November 28, 2006, p. A1.

[19] As reported by Bing West and Owen West ("Iraq's Real 'Civil War,'" *Wall Street Journal*, April 5, 2007, p. A13), AQI controlled the fuel market. Each month, ten trucks with 80,000 gallons of heavily subsidized gasoline and five trucks with kerosene were due to arrive. Instead, AQI diverted most shipments to Jordan or Syria, where prices were higher, netting $10,000 per shipment and antagonizing 30,000 shivering townspeople. No local police officer dared to make an arrest.

some tribal sheikhs allied themselves with al Qa'ida, others had decamped to Jordan and Syria.[20] The few who remained saw repeated attacks decimate their familial ranks. One such sheikh, Abdul Sattar Buzaigh al-Rishawi of the influential Rishawi tribe in Ramadi, lost his father and several brothers to AQI. In the summer of 2006, AQI killed a prominent sheikh and refused to relinquish his body for burial until days had passed, contravening Islamic tradition. Indeed, AQI posed a major threat to tribal power.

Aided by the recoil of sheikhs, Sheikh Sattar organized 25 of the 31 tribes in al Anbar to join the Anbar Salvation Council in September 2006.[21] Through the awakening movement, the sheikhs set themselves up as public enemies of AQI. Their primary strategy was to persuade young tribesmen to join the police forces of Ramadi and other al Anbar towns to help take back the province in return for protection by U.S. military forces.[22] It took several months for the alliance to build its critical strength. As promised, the sheikhs persuaded tribal members to join the local police in large numbers. By December 2006, the Ramadi police force had doubled in size from 4,000 to a plateau of 8,000.[23] In western al Anbar, the number of police went from nearly zero to 3,000.[24] The police force in the province grew to 24,000 in mid-2007, with a goal of leveling off at 30,000.[25] U.S. forces, whose policy was to

---

[20] See Greg Jaffe, "How Courting Sheiks Slowed Violence in Iraq," *Wall Street Journal*, August 8, 2007, p. A1.

[21] Khalid Al-Ansary and Ali Adeeb, "Most Tribes in Anbar Agree to Unite Against Insurgents," *New York Times*, September 18, 2006, p. A12; Edward Wong and Khalid al-Ansary, "Iraqi Sheiks Assail Cleric for Backing Qaeda," *New York Times*, November 19, 2006, p. 22; Tony Perry, "The Conflict in Iraq: Violence in the Capital," *Los Angeles Times*, January 23, 2007, p. A5.

[22] Also termed *Anbar Intifada* (Sam Dagher, "Sunni Muslim Sheikhs Join US in Fighting al Qaeda," *Christian Science Monitor*, May 3, 2007, p. 1).

[23] Tony Perry, "The Conflict in Iraq: Struggles in Al Anbar," *Los Angeles Times*, December 15, 2006, p. A17.

[24] Rick Jervis, "Police in Iraq See Jump in Recruits," *USA Today*, January 15, 2007, p. 1.

[25] al-Ansary and Adeeb (2006). Rishawi said that the 25 tribes counted 30,000 young men armed with assault rifles who were willing to confront and kill the insurgents and criminal gangs that he blamed for damaging tribal life in al Anbar.

place their officers within newly formed police units, could not staff up as quickly as police units were growing. Perhaps the most significant change was that the new police, in contrast to their predecessors, were willing to fight. As recently as August 2006, half of the police officers in Fallujah stayed home in the face of AQI threats.[26] By January, they were standing their ground.[27] A simultaneous alienation of other insurgent groups also became visible. Competing insurgent groups had seen defections deplete their ranks, and their leadership was attacked and assassinated. AQI's growing dominance of organized crime cut into the revenues on which competing groups counted. Furthermore, whereas all the major insurgent groups employed the Salafi discourse, none but AQI made the restoration of the caliphate their primary or even preferred objective.

A large share of the new police force was made up of former members of insurgent groups. One sheikh, Abu Azzam, said that the 2,300 men in his movement included members of fierce Sunni groups, such as 1920 Revolution Brigades and the Mujahideen Army, which had fought U.S. forces.[28] In late 2006, one report concluded that "elders of the Abu Soda tribe recently helped U.S. forces find IEDs [improvised explosive devices] that had been planted by their own tribesmen, and they have identified kidnappers and other local bad guys for the Americans to arrest." Sheikh Sattar added, "Even some tribes that were with the insurgency follow us."[29]

For three months, Anbar Salvation Council battled with AQI, primarily in the Ramadi area.[30] In the absence of set-piece battles, con-

---

[26] Solomon Moore and Louise Roug, "Deaths Across Iraq Show It Is a Nation of Many Wars, with U.S. in the Middle," *Los Angeles Times*, October 7, 2006, p. 1.

[27] Perry (2006).

[28] Richard A. Oppel, "Mistrust as Iraqi Troops Encounter New U.S. Allies," *New York Times*, July 16, 2007, p. A1.

[29] Sarah Childress, "Retaking Ramadi: All the Sheik's Men: U.S. Commanders Are Hoping Tribal Levies Can Help Fill the Ranks of Anbar's Police and Tackle Al Qaeda," *Newsweek*, December 18, 2006, p. 43.

[30] The course of conflict ran somewhat differently in al Anbar's far-western section (e.g., al-Qaim), its center-west section (e.g., Haditha, Ramadi), and its eastern section (e.g., Fallujah). The far-western section was already partially pacified in 2006, and, while ISI was the major

flict was conducted through constant attrition. The police rounded up those who were AQI members even as AQI extended its intimidation campaign against police, their families, and the tribal sheikhs who had turned against them. The effects of Anbar Salvation Council's formation on AQI's strategy were immediate. Prior to mid-September, fewer than 2 to 3 percent of AQI's attacks were targeted against Iraqi police. In September, the ratio started rising to reach 15 percent (this includes a few attacks on Anbar Salvation Council itself) by early December 2006, when it jumped to the 25- to 30-percent level. By contrast, whereas roughly 10 percent of AQI's attacks had targeted members of the Iraqi Army, this ratio fell in March 2007 to 5 percent, where it remained through late 2007. In effect, AQI quickly understood who was its major threat in the province: the police.

The results of this shift were dramatic. By March, al Qa'ida had largely been expelled from Ramadi, a city that had been a wasteland for U.S. and Iraqi forces.[31] Except in Fallujah, the number of attacks in al Anbar fell dramatically. U.S. deaths, which were running roughly 30 a month in the entire province, fell to three in June 2007. Sheikh al-Dulami reported that al Anbar had "been purged completely of AQI and the Anbar Salvation Council's forces are encircling the organization's remnants in al-Ta'i area south of Ramadi which makes up 0.5 percent of Anbar's overall area. AQI used to control more than 90 percent before Anbar Awakening Council's establishment last year."[32]

It may seem hard to believe that the addition of a few thousand police officers could tip the balance of power in al Anbar province so swiftly. These police officers were essentially untrained, inexperienced, and poorly equipped, especially in comparison to the 35,000 U.S. soldiers already in the province. Nevertheless, there were two critical

player there, it was not overwhelmingly dominant. The eastern section was not pacified until late 2007. It was the center-west section that underwent the largest swing in control.

[31] The assessment of Commander Steven Wisotzki, chief of staff for Naval Special Warfare Group One cited in Chris Johnson, "Military Officers Criticize Media Coverage of Battlefront in Iraq," *Inside the Navy*, February 5, 2007, p. 5.

[32] "Iraq's Sunni Armed Groups Reportedly Planning Alliance Against Al-Qa'ida," *Al-Hayay* (London), April 11, 2007.

changes. First, the new police officers had joined at the urging of their tribal leaders in a society in which tribal relationships were crucial. This bolstered their solidarity and made it more likely they would stand under pressure. Second, enough police officers knew who the individual members of AQI were and were motivated to take action against them. Since AQI drew from the same neighborhoods and demographic pool from which the police came from, this knowledge should not have been difficult to come by. But it was not exploited earlier. The police also brought a great deal of local intelligence with them into the force.[33] One U.S. Army lieutenant, Ed Clark, observed in May that "about 10 percent of our intelligence is actionable, while 90% of their intelligence is actionable."[34]

The defection of other insurgent groups was critical. The defectors had good knowledge of AQI, largely because the insurgency was networked across organizational lines prior to 2006. It was usual for Sunni insurgents not only to interact with one another but to use common sources of expertise, such as the subnetworks that created IEDs. Once the groups started breaking off from AQI, those who defected to the police force could more easily find and identify AQI leaders, thanks to their prior knowledge. In a conflict in which intelligence is paramount, former insurgent members were a crucial force multiplier.[35]

The U.S. Role in al Anbar

The United States played an important, though supportive, role. Indeed, the working relationship between the U.S. military and the Sunni tribes of al Anbar may well have made the difference between

---

[33] Chris Kraul, "In Ramadi, a Ragtag Solution with Real Results," *Los Angeles Times*, May 7, 2007, p. A6.

[34] Kraul (2007).

[35] The insurgents had, in effect, trained many of the new police officers. Consider the following newspaper story: "Marine Sgt. Tony Storey . . . recalls one man who was a natural with his AK-47. 'Where'd you learn to shoot like that?' Storey asked. 'Insurgent,' the man said with a smile" (Leila Fadel, "Iraq: Old Allegiances Loom Large as U.S. Trains Iraqi Forces," *Miami Herald*, June 17, 2007, p. A24).

failure and success, though it took years for the United States to embrace such a relationship. Success at wooing the tribes took a long time, largely because the benefits took a while to be recognized. Initially, some in the U.S. military wanted to support tribes but were told that tribal authority was a relic of Iraq's past and not part of its democratic future.[36] In late 2005, the U.S. military made a more successful attempt to gain the cooperation of tribes in the far-western part of al Anbar province, which helped pacify the area more than a year before the rest of al Anbar was pacified.[37]

The U.S. part of the al Anbar bargain was to provide protection for the sheikhs and their entourages, most visibly by parking a tank outside each sheikh's compound.[38] The sheikhs, in turn, promised to persuade tribe members to join the police forces of Ramadi and other al Anbar towns. Some reports credited U.S. Army Colonel Sean MacFarland for having convinced the al Anbar sheikhs to cooperate. He certainly pushed the project forward, but the U.S. role was more as a catalyst.[39] The sheikhs had reasons of their own to oppose AQI, which had killed many family members and cut into their operations (such as

---

[36] This was reported by Joe Klein, "Saddam's Revenge," *Time*, September 26, 2005, pp. 44–51). Military-intelligence officers presented the CPA with a plan to make a deal with 19 subtribes of the enormous Dulaimi clan, located in al Anbar province, the heart of the Sunni triangle. The tribes "had agreed to disarm and keep us informed of traffic going through their territories," says a former Army intelligence officer. "All it would have required from the CPA was formal recognition that the tribes existed—and $3 million." The money would go toward establishing tribal security forces. "It was a foot in the door, but we couldn't get the CPA to move." Bremer's spokesperson, Dan Senor, said that a significant effort was made to reach out to the tribes. But several military officials dispute that. "The standard answer we got from Bremer's people was that tribes are a vestige of the past, that they have no place in the new democratic Iraq," says the former intelligence officer.

[37] West and West (2007). As reported in late 2005, acceptably trained Iraqi battalions began to join the persistent U.S. forces in al Anbar. AQI resorted to suicide attacks and roadside bombs and avoided direct fights. Subtribes began to kill AQI members in retaliation for individual crimes and discovered that AQI was ruthless but not tough. Near the Syrian border, an entire tribe joined forces with the Marines and drove AQI from the city of al Qaim.

[38] Perry (2006). Unfortunately, such help did not prevent the assassination of Sheikh Adbul Saltar in the summer of 2007.

[39] See, for instance, Jim Michaels, "Behind Success in Ramadi: An Army Colonel's Gamble," *USA Today*, May 1, 2007a, p. A1.

smuggling) and whose concept of governance was antithetical to tribal authorities.

U.S. forces worked hard to extend what happened in al Anbar to other provinces, notably to Diyala province.[40] As one report concluded in July 2007, U.S. forces used

> Iraqi partners to pick out the insurgents and uncover the bombs they had seeded along the cratered roads [and] apprehended more than 100 militants, including several low-level emirs. . . . [L]ocal Sunni residents, including a number of former insurgents from the [1920 Revolution Brigades] . . . emerged as a linchpin of the American strategy [in Diyala].[41]

Further negotiations took place with sheikhs in Diyala.[42] In June, U.S. forces picked up pledges from support from ten tribes in Baghdad.[43] They even met with some broad success through the Sunni region as a whole.[44]

But the United States did not play a direct role. Some argued that the U.S. "surge" was pivotal.[45] But was it? The al Anbar revolt

---

[40] Gordon (2007).

[41] Gordon (2007).

[42] Trudy Rubin, "New Iraq Tribal Alliances Fighting al-Qaeda," *Philadelphia Inquirer*, June 22, 2007, p. A19. Shiite members of the tribe known as the Bani Tamim mobilized in Diyala province east of Baghdad in cooperation with Sunni tribes. This is an area where U.S. troops conducted an offensive. Leaders of the Bani Tamim had 5,000 names of tribespeople willing to fight and to protect vital oil pipelines. They wanted help from the United States to do the job.

[43] Jim Michaels, "Tribes Help U.S. Against al-Qaeda: Baghdad-Area Deals Build on Successes in Anbar Province," *USA Today*, June 20, 2007b, p. A1. More than ten Iraqi tribes in the Baghdad area have reached agreements with U.S. and Iraqi forces for the first time to oppose al Qa'ida, raising the U.S. military's hopes that a trend started in western Iraq is spreading there.

[44] John F. Burns and Alissa J. Rubin, "U.S. Arming Sunnis in Iraq to Battle Old Qaeda Allies," *New York Times*, June 11, 2007, p. A1. In an agreement announced by the U.S. command, 130 tribal sheikhs in Salahuddin met in the provincial capital, Tikrit, to form police units that would defend against al Qa'ida.

[45] Rowan Scarborough, "Petraeus Adviser: Al Qaeda Weakened by Troop Surge," *Washington Examiner*, August 15, 2007, p. 15.

against AQI started in September 2006, several months before the surge began. By mid-March 2007, when the conversion of al Anbar was largely complete, most of the additional forces promised to al Anbar had yet to arrive. Others argued that the United States had to show staying power to persuade the sheikhs to counter AQI. Yet sentiment in the United States in favor of staying was ebbing in the fall of 2006, when Anbar Salvation Council was formed and was further weakened by the turnover of U.S. House and Senate leadership as a consequence of the November 2006 elections.[46] As one U.S. Marine noted, "For three years we fought our asses off out here and made very little progress. Now we are working with the sheikhs, and Ramadi has gone from the most dangerous city in the world to a place where I can sit on Sheik [Hamid] Heiss's front porch without my body armor and not have to worry about getting shot."[47] Some credit an offensive carried out by Marines that same month.[48] Yet, at the time, it looked no different from many other offensives that took place in earlier years. Others credit offensives carried out by U.S. forces in late February and early March.[49] They probably helped on the margins, but by then the momentum from operations carried out by newly recruited police were more decisive.

---

[46] To quote Colonel MacFarland, one of the Americans responsible for the Awakening,

> A growing concern that the U.S. would leave Iraq and leave the Sunnis defenseless against al-Qaeda and Iranian-supported militias made these younger [tribal] leaders [who led the Awakening] open to our overtures. (Major Neil Smith and Colonel Sean MacFarland, "Anbar Awakens: The Tipping Point," *Military Review*, March–Aprril 2008, p. 51)

[47] Jaffe (2007).

[48] John F. Burns, "Showcase and Chimera in the Desert," *New York Times*, July 8, 2007, p. A1. The problem with the Burns quote, "the sheiks turned only after a prolonged offensive by American and Iraqi forces, starting in November, that put Al Qaeda groups on the run, in Ramadi and elsewhere across western Anbar," is that Anbar Salvation Council was actually formed in September, or two months earlier.

[49] As reported by Max Boot, "Can Petraeus Pull It Off?" *Weekly Standard*, April 30, 2007, p. 24. The principal operations were code-named Murfreesboro (February 10–March 10), Okinawa (March 9–20), and Call to Freedom (March 17–30).

The Logic of Military Force

Policing and politics may be preferable against most terrorist groups. But military force may sometimes be more useful against large terrorist groups engaged in insurgencies. In Iraq, military force was helpful, at least temporarily, against AQI. But the U.S. military largely played a background role. AQI subverted many of the governing institutions, acquired lucrative taps into the local economy (notably oil distribution), and had started to impose its own version of shari'a law where it could. Nationwide, AQI made determined efforts to create a national front, with itself as the vanguard of the Sunni movement. This was done first in January 2006 with the formation of MSC and later in October 2006 with the formation of ISI. In that role, it assimilated several terrorist-insurgent groups, peeled off brigades from other insurgent groups, and pressured the rest to join ISI. Those who refused became targets of intimidation.

In essence, AQI pushed others to get on the bandwagon, with some success. The formation of Anbar Salvation Council was largely a backlash to the tactics and techniques used by AQI against those tribes and insurgent terrorist groups that refused to join its ranks. For several months, Anbar Salvation Council, under pressure from AQI, worked without apparent success. It took concerted effort over months to move tribe members into the local police forces, and many of the new Iraqi police were former insurgents themselves. By joining the local government, they deserted the insurgency.

To examine more systematically what works against terrorist groups that have graduated into becoming insurgencies, we constructed a list of 84 insurgent groups that were also terrorist groups and analyzed the results.[50] Terrorist groups that are involved in insurgencies differ sharply from other terrorist groups. Of these 84, only two were

---

[50] If an insurgent group used terror, we coded it as a terrorist group. Most terrorist groups, however, are not insurgent groups, because they are not powerful enough to take on a government's military forces and mount effective resistance. We used data from Fearon and Laitin's (2003) work on insurgencies and civil wars. The additional insurgencies were those that met the Fearon-Laitin criteria after 1999, the data cutoff point for their work.

associated with high-income countries: Northern Ireland and Greece.[51] By contrast, almost a third of terrorist groups (173) are located in rich countries. Table 5.1 illustrates this breakdown in finer detail. In essence, the poorer the country, the higher a percentage of terrorist groups that reach insurgency stage. Whether this is a phenomenon of income or represents the underreporting of smaller, weaker terrorist groups is, however, difficult to determine. Table 5.1 highlights the results.

Terrorist groups involved in insurgencies tend to be larger than most terrorist groups, since size is a key prerequisite for seizing power. Here, the breakpoint seems to be 1,000 members, as Table 5.2 indicates. Larger than that and the group is more likely to be involved in an insurgency; smaller than that and it is rarely involved in an insurgency.

Terrorist groups involved in insurgencies tend to focus on mid-level goals, as Table 5.3 illustrates. They rarely seek broad goals, such as social revolution or empire. And they rarely seek narrow goals, such as the status quo or policy change. Rather, they tend to seek the overthrow of regimes or territorial change, especially secession. This suggests that there may be some negotiating room with insurgent groups. As noted earlier, the possibility of a political solution is linked to a key variable: the breadth of goals. The narrower a group's goals, the more willing a government may be to negotiate with it.

Table 5.1
Insurgent Groups and Income

Income	Insurgents	Others	Insurgent Share (%)
High income	2	182	1
Upper middle income	9	84	10
Lower middle income	38	190	17
Low income	35	108	24

[51] The Basque group Basque Fatherland and Freedom (Euskadi Ta Askatasuna, or ETA) is not included, because the death toll associated with its efforts fell just short of the 1,000 required for inclusion as an insurgency.

Table 5.2
Insurgent Groups and Size

Size of Group	Insurgents	Others	Insurgent Share (%)
>10,000 members	19	11	63
1,000–9,999 members	45	37	55
100–999 members	13	157	8
0–99 members	7	359	2

Table 5.3
Insurgent Groups and Goals

Group Goal	Insurgents	Others	Insurgent Share (%)
Social revolution	0	75	0
Empire	2	22	8
Regime change	44	177	20
Territorial change	30	146	17
Policy change	5	123	4
Status quo	3	21	12

The final factor is how insurgent groups end. The results are shown in Table 5.4. When a terrorist group becomes an insurgent group, it does not go easily. Half of the insurgent groups have not ended (seven of these were Iraqi, four were Palestinian). The data show that, when insurgent groups have ended, nearly half of the time, they negotiated a settlement with the government. A quarter of the time, the group achieved victory. If a political solution is not feasible and the group did not achieve victory, military force has often been a viable option. Roughly 19 percent of insurgent groups ended because military forces defeated them. Policing is rarely effective against insurgent groups when used as the primary tool, since well-armed, well-motivated groups tend to overmatch police.

Table 5.4
The End of Insurgent Groups

End of Group	Insurgents	Others
Splintering	4	132
Policing	2	105
Victory	10	17
Military force	8	12
Politics	18	96

CHAPTER SIX

The Limits of America's al Qa'ida Strategy

So far, this book has examined how terrorist groups end. The rest of the book turns to implications for dealing with al Qa'ida based on this analysis. This chapter argues that al Qa'ida has been involved in more terrorist attacks in a wider geographical area since September 11, 2001, than it had been during its previous history. These attacks spanned Europe, Asia, the Middle East, and Africa. Al Qa'ida's modus operandi evolved and included a repertoire of more-sophisticated IEDs and a growing use of suicide attacks. Its organizational structure has also evolved, making it a more dangerous enemy. This includes a bottom-up approach (encouraging independent thought and action from low-level operatives) and a top-down one (issuing orders and still coordinating a far-flung terrorist enterprise with both highly synchronized and autonomous moving parts).[1]

There has been some discrepancy about the effectiveness of U.S. strategy against al Qa'ida. In 2007, for example, vice president Dick Cheney stated that the United States had "struck major blows against the al-Qaeda network that hit America."[2] Pakistan's president Pervez Musharraf claimed that "Pakistan has shattered the al Qa'ida network in the region, severing its lateral and vertical linkages. It is now on the run and has ceased to exist as a homogenous force, capable of under-

---

[1] See, for example, Hoffman (2006, pp. 285–289).

[2] Dick Cheney, "Vice Presidents Remarks at a Rally for the Troops," USS *John C. Stennis*, May 11, 2007.

taking coordinated operations."[3] The *National Security Strategy of the United States* boldly stated, "The al-Qaida network has been significantly degraded."[4] These arguments were regularly repeated after 2001. "Al Qaeda's Top Primed to Collapse, U.S. Says," read a *Washington Post* headline[5] two weeks after Khalid Sheikh Mohammed, the mastermind behind the September 11 attacks, was arrested in March 2003.[6]

Our analysis suggests that these claims were overstated. A growing body of work supports our conclusion. For example, the 2008 *Annual Threat Assessment of the Director of National Intelligence* reported that, "Using the sanctuary in the border area of Pakistan, al-Qa'ida has been able to maintain a cadre of skilled lieutenants capable of directing the organization's operations around the world." It also noted that "Al-Qa'ida is improving the last key aspect of its ability to attack the US: the identification, training, and positioning of operatives for an attack in the Homeland."[7] The 2007 national intelligence estimate, *The Terrorist Threat to the US Homeland*, similarly noted that the main threat to the U.S. homeland "comes from Islamic terrorist groups and cells, especially al-Qa'ida, driven by their undiminished intent to attack the Homeland and a continued effort by these terrorist groups to adapt and improve their capabilities."[8] Bruce Riedel, who spent 29 years at the CIA, acknowledged that "Al Qaeda is a more dangerous enemy today than it has ever been before."[9]

---

[3] Pervez Musharraf, *In the Line of Fire: A Memoir*, New York: Free Press, 2006, pp. 273–274.

[4] George W. Bush, *The National Security Strategy of the United States of America*, Washington, D.C.: White House, 2006, p. 8.

[5] Dana Priest and Susan Schmidt, "Al Qaeda's Top Primed to Collapse, U.S. Says: Mohammed's Arrest, Data Breed Optimism," *Washington Post*, March 16, 2003, p. A1.

[6] Priest and Schmidt (2003).

[7] J. Michael McConnell, *Annual Threat Assessment of the Director of National Intelligence for the Senate Select Committee on Intelligence*, February 5, 2008, p. 6.

[8] Office of the Director of National Intelligence and National Intelligence Council, *The Terrorist Threat to the US Homeland*, Washington, D.C., 2007, p. 5.

[9] Bruce Riedel, "Al Qaeda Strikes Back," *Foreign Affairs*, Vol. 86, No. 3, May–June 2007, p. 24.

Based on our assessment, the United States should fundamentally rethink its strategy toward al Qa'ida. U.S. efforts have relied too heavily on military force. This chapter is divided into five sections. First, it examines—and critiques—the current U.S. approach, which relies predominantly on military force to target al Qa'ida around the globe. Second, it examines al Qa'ida's attacks and geographic scope since 1995. Third, it assesses al Qa'ida's organizational structure. Fourth, it outlines central al Qa'ida's base of operations in Pakistan and Afghanistan. Fifth, it offers some brief conclusions.

The Limits of Military Force

The current U.S. strategy against al Qa'ida centers on the use of military force. Military force was not, of course, the only instrument that the United States used against al Qa'ida. The U.S. Department of State engaged in a range of diplomatic counterterrorism initiatives, including through its Antiterrorism Assistance Program. The FBI and local police agencies historically tracked and arrested terrorists in the United States. The U.S. Department of Homeland Security implemented numerous policies at ports of entry and critical infrastructure to secure the United States from terrorist attacks. The U.S. Department of Treasury targeted terrorist financial networks. And the CIA collected, analyzed, and conducted operations against terrorist groups abroad. While a range of instruments was used, the military component was paramount in two respects.

First, U.S. policymakers and key national-security documents referred to operations against al Qa'ida as the *global war on terror*.[10] The use of the word *war* to describe U.S. efforts had an important symbolic importance, since it suggested a military conflict that required a military solution. As the U.S. Department of Defense's 2006 *Quadrennial Defense Review* noted, "The United States is a nation engaged in

[10] On the Bush administration debates, see Douglas J. Feith, *War and Decision: Inside the Pentagon at the Dawn of the War on Terrorism*, New York: HarperCollins Publishers, 2008, pp. 47–87.

what will be a long war. Since the attacks of September 11, 2001, our Nation has fought a global war against violent extremists who use terrorism as their weapon of choice, and who seek to destroy our free way of life."[11] To win, the document concluded, the United States needed to make U.S. military forces more agile and more expeditionary. This included making technological advances, such as dramatic improvements in information management and precision weaponry, as well as changing U.S. global military-force posture. This strategy was consistent with other U.S. documents, such as *The National Security Strategy of the United States of America*, which argued, "America is at war. This is a wartime national security strategy required by the grave challenge we face—the rise of terrorism fueled by an aggressive ideology of hatred and murder, fully revealed to the American people on September 11, 2001."[12]

Second, the U.S. military spent the bulk of U.S. counterterrorism resources. With the high cost of military platforms and systems, military budgets were bound to be larger than those for other departments. This is partly why the U.S. Department of Defense's fiscal-year 2008 budget was six times larger than the combined budgets of the U.S. Department of Justice, U.S. Department of State, U.S. Department of Homeland Security, U.S. Department of Treasury, U.S. Agency for International Development (USAID), and Office of the Director of National Intelligence.[13] But funding devoted to counterterrorism can nonetheless provide a rough estimate of priorities.

---

[11] U.S. Department of Defense, *Quadrennial Defense Review Report*, Washington, D.C., 2006, p. v.

[12] Bush (2006, p. i).

[13] Their fiscal-year 2008 budgets were as follows: U.S. Department of Defense, $623.1 billion; U.S. Department of Homeland Security, $34.3 billion; U.S. Department of Justice, $20.2 billion; U.S. Department of the Treasury, $12.1 billion; U.S. Department of State, $35 billion; national intelligence program, $43.5 billion. These numbers include their discretionary budgets, supplemental amounts for the global war on terrorism, and supplemental amounts for homeland security (Office of Management and Budget, *The Budget of the United States Government*, Washington, D.C.: Executive Office of the President, Office of Management and Budget, last updated January 28, 2008).

After September 11, 2001, increases in annual U.S. Department of Defense spending dwarfed increases in spending for all other departments critical to counterterrorism combined (U.S. Department of Justice, U.S. Department of State, and U.S. Department of Homeland Security) by five to one, even when the costs of the wars in Afghanistan and Iraq were excluded.[14] In aggregate terms, the Pentagon significantly outspent other U.S. government agencies on counterterrorism-related issues. Between 2001 and 2007, for example, the U.S. Congress approved a total of $609 billion for a wide range of counterterrorism purposes. Approximately 90 percent went to the U.S. Department of Defense. These resources went to military operations and assistance in a range of areas, such as Iraq, Afghanistan, Pakistan, Philippines, Somalia, and the Horn of Africa. The U.S. Department of State and USAID together received about $40 billion for counterterrorism assistance and foreign-aid programs over this period.[15] As one assessment of U.S. counterterrorism spending concluded,

> As we ask our military to become the leading edge of our international engagement, we are putting a security face on that engagement. However benign and well-intended our forces, for other nations and peoples this can create a backlash against our policies and our presence. In the end, leading with our military chin could have the effect of endangering, rather than increasing, American security.[16]

Military force is too blunt an instrument to defeat most terrorist groups. Military forces may be able to penetrate and garrison an area that terrorist groups frequent and, if well sustained, may temporar-

---

[14] David C. Gompert and James Dobbins, "A Far Too Costly Pentagon," United Press International, February 27, 2006.

[15] Amy Belasco, *The Cost of Iraq, Afghanistan, and Other Global War on Terror Operations Since 9/11*, Washington, D.C.: Congressional Research Service, Library of Congress, 2007, p. 6. On the cost of the global war on terrorism, also see Steven M. Kosiak, *The Global War on Terror (GWOT): Costs, Cost Growth and Estimating Funding Requirements*, testimony before the U.S. Senate Committee on Budget, February 6, 2007.

[16] Gordon Adams, *Budgeting for Iraq and the GWOT: Testimony, Committee on the Budget*, United States Senate, February 6, 2007.

ily reduce terrorist activity. But once the situation in an area becomes untenable for terrorists, they will simply transfer their activity to another area, and the problem remains unresolved.[17] Terrorist groups generally fight wars of the weak. They do not put large, organized forces into the field, except when they engage in insurgencies.[18] This means that military forces can rarely engage terrorist groups using what most armies are trained in: conventional tactics, techniques, and procedures. Most soldiers are not trained to understand, penetrate, and destroy terrorist organizations. Former secretary of defense Donald Rumsfeld ironically noted in a private memo in 2003 that U.S. Department of Defense capabilities were not adequate to defeat terrorist organizations: "DoD has been organized, trained and equipped to fight big armies, navies and air forces. It is not possible to change DoD fast enough to successfully fight the global war on terror."[19]

On the one hand, military tools have increased in precision and lethality, especially with the growing use of standoff weapons and imagery to monitor terrorist movement. These capabilities may limit the footprint of U.S. or other forces and minimize the costs and risks of sending in military forces to potentially hostile countries.[20] On the other hand, even precision weapons have been of marginal use against terrorist groups. For example, the United States launched cruise-missile strikes against facilities in Afghanistan and Sudan in response to the 1998 bombing of U.S. embassies in Tanzania and Kenya. But they had no discernible impact on al Qa'ida. The use of massive military power against terrorist groups also runs a significant risk of turning the population against the government. This terrorist strategy is often referred to as *provocation*: Terrorist groups seek to goad the target government into a military response that harms civilians within the terrorist organization's home territory. The aim is to convince the population that

---

[17] David Galula, *Counterinsurgency Warfare: Theory and Practice*, New York: Praeger, 1964, p. 72.

[18] See, for example, Luttwak (2001).

[19] Rumsfeld (2003).

[20] Pillar (2001, pp. 97–110); Hoyt (2004).

the military is so evil that the terrorists' radical goals are justified.[21] Al Qa'ida figures, including Ayman al-Zawahiri, adopted the strategy in such countries as Egypt "to force the Egyptian regime to become even more repressive, to make the people hate it."[22]

The use of military force in Iraq was a specific problem. The United States diverted precious resources and scarce attention to overthrowing Saddam Hussein's government and away from the war against al Qa'ida in most of the world. Gary Schroen, who led the first CIA team into Afghanistan in 2001, argued that the war in Iraq drained Afghanistan of key U.S. military personnel. Iraq also drained key CIA personnel and resources from Afghanistan, "making it increasingly difficult to staff the CIA teams in Afghanistan with experienced paramilitary officers."[23] Indeed, the United States did not have a sufficient number of personnel to target key al Qa'ida leaders in Pakistan and other countries, because they were diverted to Iraq. The capture of many key al Qa'ida leaders, such as Khalid Sheikh Mohammed, Ramzi Binalshibh, and Abu Zubeida, occurred between 2001 and 2003. After the overthrow of Saddam's government, this attention shifted to stabilizing an increasingly violent war in Iraq. In sum, the United States lacked the resources and attention to adequately defeat al Qa'ida, because it was diverted to Iraq.

Iraq was also helpful to al Qa'ida, since it established a foothold that it did not previously have. On the occasion of the second and third anniversaries of the September 11 attacks, the group's second-in-command, Ayman al-Zawahiri provided the clearest explanation of al Qa'ida's strategy in Iraq: He declared in September 2003,

> We thank God for appeasing us with the dilemmas in Iraq and Afghanistan. The Americans are facing a delicate situation in

[21] Kydd and Walter (2006, pp. 69–72); David Fromkin, "The Strategy of Terrorism," *Foreign Affairs*, Vol. 53, No. 4, July 1975, pp. 683–698.

[22] Lawrence Wright, *The Looming Tower: Al-Qaeda and the Road to 9/11*, New York: Knopf, 2006, p. 217.

[23] Gary C. Schroen, *First In: An Insider's Account of How the CIA Spearheaded the War on Terror in Afghanistan*, New York: Presidio Press/Ballantine Books, 2005, p. 360.

> both countries. If they withdraw they will lose everything and if they stay, they will continue to bleed to death.[24]

Such images as the Abu Ghraib prisoners were sent around the globe via Internet, satellite television, and cell phone. The war in Iraq also created a perception that Islam was under threat. Many Muslims accepted al Qa'ida's argument that jihad was justified precisely because Islam was under attack by the United States. Consequently, fighting ground wars in the Muslim world appeared to inflame, not quell, Islamic terrorism.

The Return of al Qa'ida

Indeed, the evidence since September 11, 2001, strongly suggests that the U.S. strategy was not successful in undermining al Qa'ida's capabilities in the long run. Al Qa'ida remained a strong and competent organization. Its goals were the same: uniting Muslims to fight the United States and its allies (the far enemy) and overthrowing west-friendly regimes in the Middle East (the near enemy) to establish a pan-Islamic caliphate.[25]

Al Qa'ida was involved in more terrorist attacks in the first six years after September 11, 2001, than it had been during the previous six years. It averaged fewer than two attacks per year between 1995 and 2001, but it averaged more than ten attacks per year between 2002 and 2007. Figure 6.1 includes attacks in which al Qa'ida was directly involved between 1995 and 2007. The database is available in Appendix A. We excluded attacks in Iraq and Afghanistan. Since al Qa'ida attacks were part of a much broader insurgency, we found it difficult to disentangle which attacks al Qa'ida perpetrated and which other

---

[24] Michael Scheuer, *Imperial Hubris: Why the West Is Losing the War on Terror*, Washington, D.C.: Brassey's, 2004, p. xxi.

[25] On the establishment of a caliphate, see, for example, Abu Bakr Naji, *The Management of Savagery: The Most Critical Stage Through Which the Umma Will Pass*, Cambridge, Mass.: John M. Olin Institute for Strategic Studies, Harvard University, 2006.

Figure 6.1
Al Qa'ida Attacks, 1995–2007 (excluding Iraq and Afghanistan)

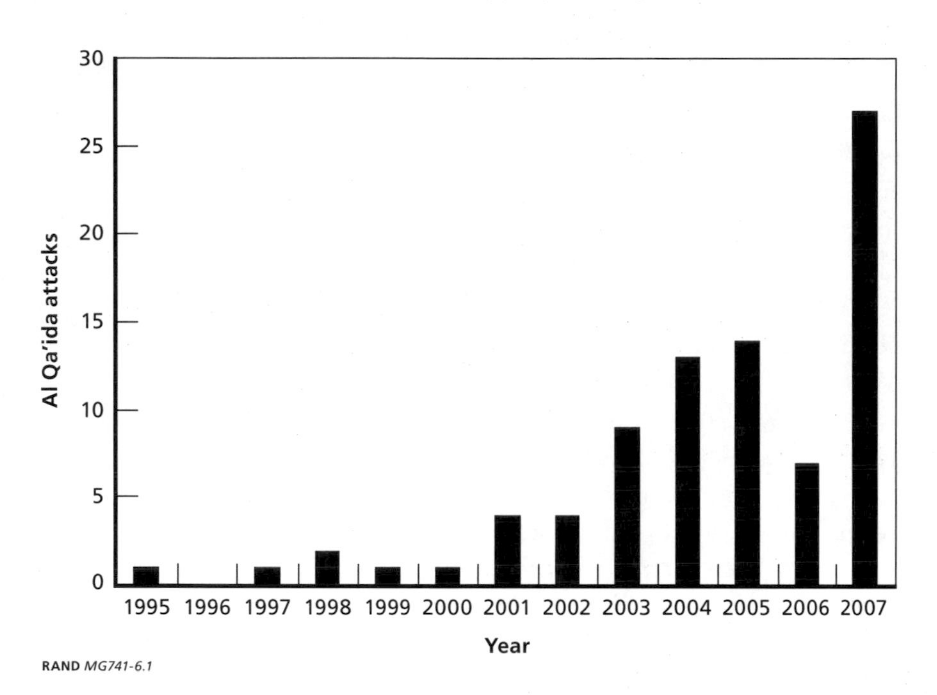

RAND *MG741-6.1*

groups perpetrated.[26] For example, there are no reliable estimates—and no way to reliably assess—which attacks in Afghanistan included a significant al Qa'ida component. Several organizations—such as the Taliban, al Qa'ida, and the Haqqani network—were involved in suicide and IED attacks. But there was no dependable way to identify which ones involved al Qa'ida. Nonetheless, al Qa'ida's direct role in the Afghanistan and Iraq insurgencies, both of which occurred after September 11, 2001, strengthens the argument that it was involved in more attacks in the first six years after September 11, 2001, than it was before that date.

Between 1995 and 2001, al Qa'ida was involved in a range of terrorist attacks beginning with the November 13, 1995, attack in Riyadh, Saudi Arabia. The explosion occurred outside the office of the program manager of the Saudi Arabian National Guard and killed seven people:

---

[26] This task is especially difficult using unclassified sources.

five Americans and two Indian government officials. The Saudi government arrested four perpetrators, who admitted connections with bin Laden.[27] Over the subsequent several years, al Qa'ida conducted several attacks in such locations as Kenya, Tanzania, and Yemen. In 1996, bin Laden issued a declaration of jihad against the United States and noted that the "presence of the American Crusader military forces on land, sea and air of the states of the Islamic Gulf is the greatest danger threatening the largest oil reserve in the world." Consequently, he said, U.S. oppression of the Holy Land "cannot be demolished except in a rain of bullets."[28] In 1998, bin Laden called specifically for the murder of any American, anywhere on earth, as the "individual duty for every Muslim who can do it in any country in which it is possible to do it."[29] Even within such European countries as the United Kingdom, a growing number of extremists sought training from al Qa'ida in Pakistan and Afghanistan. A British House of Commons report noted that, as "Al Qa'ida developed in the 1990s, a number of extremists in the UK, both British and foreign nationals—many of the latter having fled from conflict elsewhere or repressive regimes—who began to work in support of its agenda, in particular, radicalising and encouraging young men to support jihad overseas."[30]

After 2001, al Qa'ida significantly increased its number of attacks, which spanned a wider geographic area across Europe, Asia, the Middle East, and Africa. As Figure 6.2 indicates, al Qa'ida continued to conduct attacks in several key locations, such as Saudi Arabia and Kenya. But it also expanded into North Africa (Tunisia and Algeria), Asia (Bangladesh, Indonesia, and Pakistan), the Middle East (Jordan,

---

[27] National Commission on Terrorist Attacks upon the United States, *The 9/11 Commission Report: Final Report of the National Commission on Terrorist Attacks upon the United States*, Washington, D.C., 2004, p. 60; Daniel Benjamin and Steven Simon, *The Age of Sacred Terror*, New York: Random House, 2002, pp. 132, 242.

[28] Osama bin Laden, "Declaration of War Against the Americans Occupying the Land of the Two Holy Places," August 23, 1996.

[29] Osama bin Laden, "World Islamic Front for Jihad Against Jews and Crusaders: Initial 'Fatwa' Statement," *al-Quds al-Arabi*, February 23, 1998, p. 3.

[30] House of Commons, *Report of the Official Account of the Bombings in London on 7th July 2005*, London: The Stationery Office, HC 1087, 2006, p. 29.

Figure 6.2
Al Qa'ida's Geographic Expanse

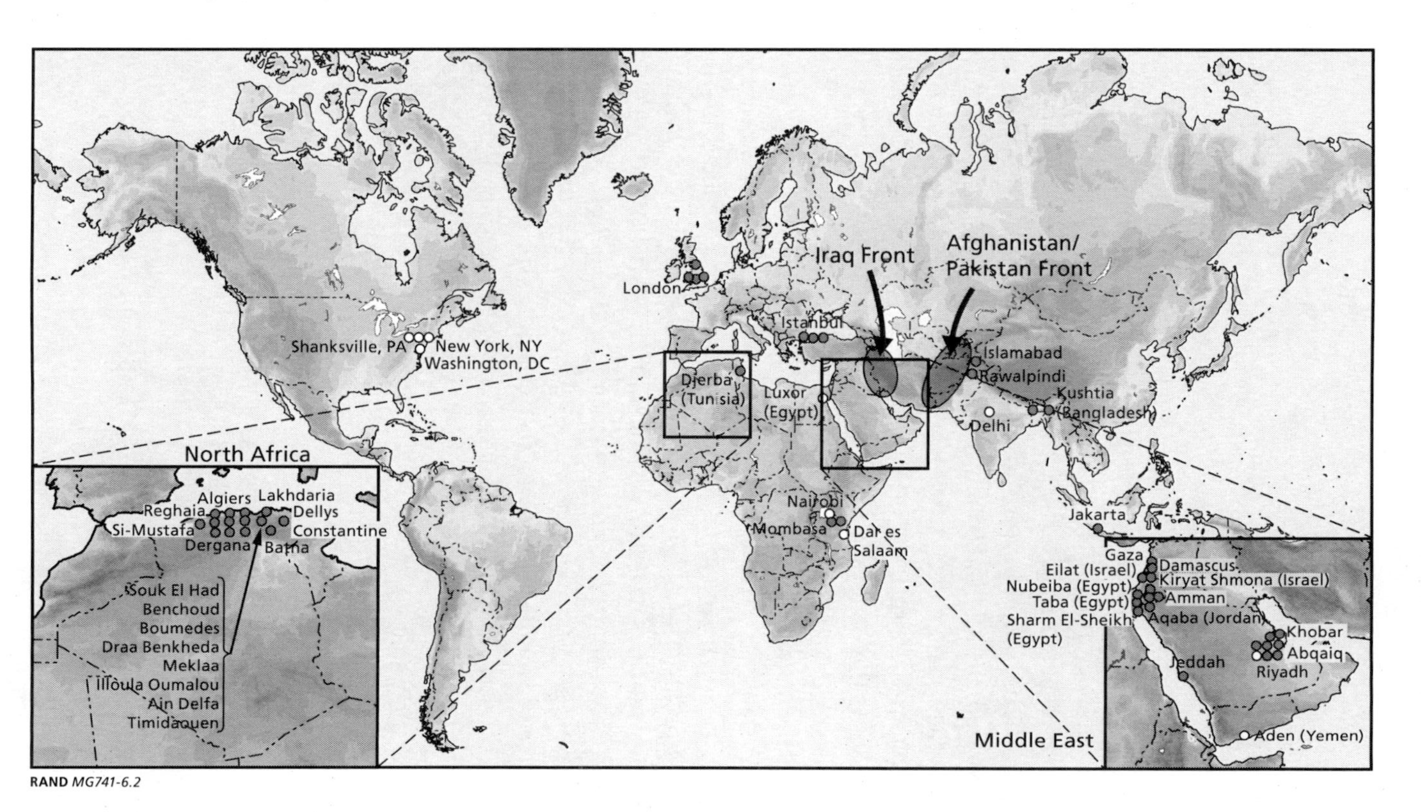

RAND *MG741-6.2*

Turkey, and Egypt), and Europe (the United Kingdom). Most of these attacks were located in the area controlled by the caliphate, notably the Umayyad caliphate from 661 to 750 AD. This was part of al Qa'ida's visionary pan-Islamic caliphate.[31]

Al Qa'ida also became involved in two major insurgencies against U.S. forces. The first was in Afghanistan, where it assisted the Taliban, Gulbuddin Hekmatyar's Hezb-i-Islami, Jalaluddin Haqqani's network, and a variety of other organizations in their struggle against Hamid Karzai's government. The second was in Iraq.

Public-opinion polls also showed notable support for al Qa'ida. In a poll released in 2007 by the University of Maryland's Program on International Policy Attitudes, for example, 25 percent of Egyptians interviewed said that they supported al Qa'ida's attacks on Americans and shared its attitudes toward the United States. Another 31 percent of Egyptians opposed al Qa'ida's attacks on Americans but shared many of its attitudes toward the United States. Furthermore, 40 percent of Egyptians, 27 percent of Moroccans, 27 percent of Pakistanis, and 21 percent of Indonesians had positive feelings toward Osama bin Laden.[32] Support for al Qa'ida also declined in several countries. In Jordan, for instance, 55 percent of those interviewed in 2003 said that they had either a lot or some confidence in bin Laden. This jumped to 60 percent in 2005 but declined to 25 percent in 2006, perhaps because of the November 2005 attacks on the Grand Hyatt hotel in Amman, which were linked to al Qa'ida. In Pakistan, 45 percent said that they had either a lot or some confidence in bin Laden. This increased to 51 percent in 2005 but then dipped to 38 percent in 2006.[33] The results of these polls should be taken with great caution, since verbal support for al Qa'ida or bin Laden does not necessarily translate into action.

---

[31] See, for example, Bernard Lewis, *The Crisis of Islam: Holy War and Unholy Terror*, New York: Modern Library, 2003, p. xi.

[32] Steven Kull, Clay Ramsay, Stephen Weber, Evan Lewis, Ebrahim Mohseni, Mary Speck, Melanie Ciolek, and Melinda Brouwer, *Muslim Public Opinion on US Policy, Attacks on Civilians and al Qaeda*, College Park, Md.: WorldPublicOpinion.org, Program on International Policy Attitudes, University of Maryland, 2007.

[33] Pew Research Center, *The Great Divide: How Westerners and Muslims View Each Other*, Washington, D.C., 2006.

Nonetheless, al Qa'ida retained a notable network of sympathizers. More than 25 percent of those interviewed in a number of Muslim countries—such as Indonesia, Pakistan, Jordan, and Nigeria—had either a lot or some confidence in bin Laden.

Adaptive Organizational Structure

As Bruce Hoffman posited, the al Qa'ida movement could most usefully be conceptualized as comprising four dimensions: al Qa'ida central, affiliated groups, affiliated units, and al Qa'ida's informal network.[34]

Al Qa'ida central was based in Pakistan and included both old and new faces, despite the death or capture of key al Qa'ida figures, such as Khaled Sheikh Mohammed, Abu Faraj al-Libi, Abu Hamza Rabia, and Abd al-Hadi al-Iraqi.[35] They included Osama bin Laden, Ayman al-Zawahiri, and Mustafa Ahmed Muhammad Uthman Abu al-Yazid. Most of al Qa'ida central was based out of the Afghanistan-Pakistan border area, especially in Pakistan's Federally Administered Tribal Areas. In some cases, this included direct command and control. In other cases, operatives made contact with, and enlisted the assistance of, local sympathizers.[36]

The *affiliated groups* included formally established terrorist groups that benefited from bin Laden's guidance and received training, arms, money, or other assistance from al Qa'ida. Among the recipients of this assistance were terrorist groups and insurgent forces, such as al Qa'ida in Iraq, Islamic Movement of Uzbekistan, and some of the Kashmiri Islamic groups based in Pakistan, such as JeM and Lashkar-e-Taiba.[37]

[34] This section adopts the framework laid out by Bruce Hoffman (2006, pp. 285–289). Also see Sageman (2004) and Brian A. Jackson, "Groups, Networks, or Movements: A Command-and-Control–Driven Approach to Classifying Terrorist Organizations and Its Application to Al Qaeda," *Studies in Conflict and Terrorism*, Vol. 29, No. 3, May 2006, pp. 241–262.

[35] Indeed, six months after September 11, 2006, al Qa'ida lost 16 of 25 key leaders (Gunaratna, 2002, p. 303).

[36] Hoffman (2006, pp. 285–289).

[37] Hoffman (2007).

Indeed, the Salafist Group for Preaching and Combat (Groupe Salafiste pour la Prédication et le Combat, or GSPC) officially merged with al Qa'ida in September 2006, subsequently changed its name to al-Qa'ida in the Islamic Maghreb and attacked a U.S. contractor bus in December 2006 in greater Algiers, marking its first attack against a U.S. entity.[38]

The *affiliated units* were loosely knit cells of radicals who had some direct connection with al Qa'ida. Unlike the previous category, these units were not large insurgent organizations attempting to overthrow their local governments. Some had prior terrorism experience in such campaigns as Algeria, the Balkans, Chechnya, Afghanistan, or Iraq and may have trained in an al Qa'ida facility before September 2001. Others had no prior experience, such as the individuals involved in the July 2005 bombing in London and led by Mohammad Sidique Khan. In all of these cases, the terrorists had some direct link with al Qa'ida.[39]

Finally, the broader *al Qa'ida network* included Islamic radicals with no direct connection with al Qa'ida but were still willing to conduct attacks in support of al Qa'ida's jihadi agenda. They were motivated by hatred toward the United States and the West, as well as toward allied regimes in the Middle East, such as Egypt and Saudi Arabia. However, the relationship with al Qa'ida was inspirational.[40] One example was the four men charged in June 2007 with plotting to blow up fuel tanks, terminal buildings, and the web of fuel lines running beneath John F. Kennedy International Airport (which the plotters code-named "chicken farm"). They had no direct connections with al Qa'ida but were prepared to conduct attacks in solidarity with it.[41]

[38] DoS (2007, p. 269).

[39] Hoffman (2006, pp. 285–289).

[40] Hoffman (2006, pp. 285–289).

[41] New York City Police Department, *Threat Analysis: Subject: JFK Airport/Pipeline Plot*, New York, June 2, 2007.

Al Qa'ida Central

Al Qa'ida central posed the most significant threat to the United States. As Figure 6.3 illustrates, much of al Qa'ida's presence was located in an area that began around the Bajaur tribal agency in Pakistan and Konar province in Afghanistan and snaked southward along the border to South Waziristan in Pakistan and Paktika province in Afghanistan. In this area, al Qa'ida was involved in an array of activities, such as strategic planning, information operations, training, and bomb-making for its global terrorist efforts and its more localized efforts in Afghanistan and Pakistan.

On the international front, there was significant evidence of al Qa'ida involvement in high-profile international attacks from its base in Pakistan and Afghanistan. The connection between affiliated units and central al Qa'ida elements can be critical. Central al Qa'ida can provide inspiration, technical expertise (such as the use of hydrogen peroxide), money, and overall guidance. The 2005 terrorist attack in

Figure 6.3
Al Qa'ida's Base of Support, 2008

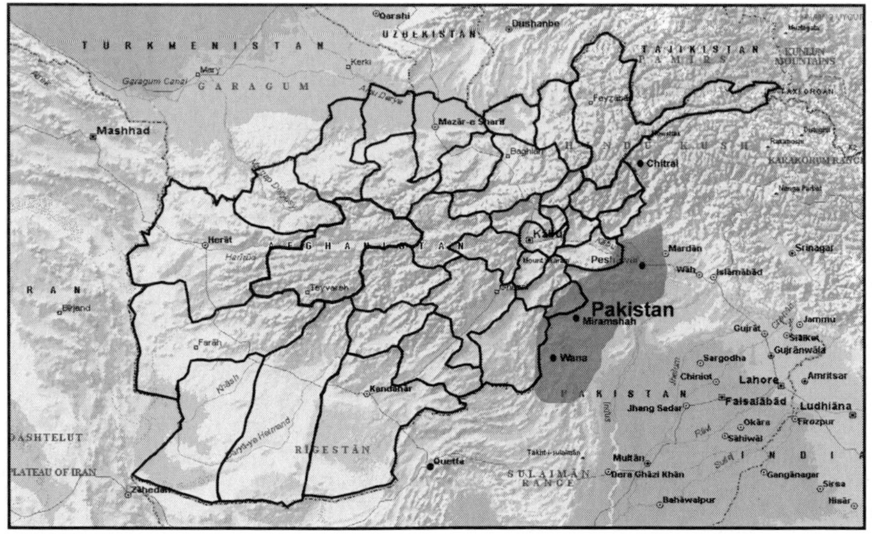

RAND *MG741-6.3*

London is a good example. Mohammad Sidique Khan and a fellow bomber, Shahzad Tanweer, visited Pakistani terrorist camps between November 2004 and February 2005, where al Qa'ida operatives trained them.[42]

Al Qa'ida was involved in the 2006 transatlantic plot to bomb U.S. airliners and crash them into targets in U.S. cities, what the British referred to as Operation Overt.[43] There were also other links to al Qa'ida and other terrorist groups in Pakistan, including the plot to attack U.S. and German targets in 2007 and the 2008 plot to attack targets in Spain and France.

In addition to its international component, al Qa'ida was also directly involved in the Afghanistan and Pakistan insurgencies, where it acted as a force multiplier for the Taliban and allied commanders, such as Sirajuddin Haqqani and Baitullah Mehsud. In 2007, for example, al Qa'ida appointed Mustafa Abu al-Yazid to head its insurgency operations in Afghanistan.[44] Al Qa'ida cooperated with these groups at the strategic level, helping develop and implement campaign strategy. It helped improve their information operations, especially the use of the Internet and video. It created a variety of Web sites, such as Voice of Jihad,[45] and used al Qa'ida's production company, as-Sahab Media, to make videos. Indeed, the Taliban's strategic information campaign significantly improved since September 11, 2001, thanks in part to al Qa'ida. The Taliban's videos were notably better in video quality and clarity of message, and its use of the Internet dramatically increased to spread propaganda and recruit potential fighters.

Al Qa'ida was also involved at the operational and tactical levels, helping Afghan groups with suicide tactics and IEDs. The number of suicide attacks increased from one in 2001 and 2002 to two in 2003,

---

[42] House of Commons (2006, pp. 20–21).

[43] On the plot's connection to al Qa'ida, see U.S. Department of State, Office of the Coordinator for Counterterrorism, "NCTC Observations Related to Terrorist Incidents Statistical Material," *Country Reports on Terrorism 2006*, Washington, D.C., 2007, p. 269.

[44] "Al Jazeera Reveals New al Qaeda Leader," *Washington Times*, May 25, 2007, p. 17.

[45] Voice of Jihad, "Imarat Islami of Afghanistan," undated Web page.

six in 2004, 21 in 2005, 139 in 2006, and 140 in 2007.[46] Al Qa'ida was involved in recruiting suicide bombers from such places as Afghan refugee camps in Pakistan; training them at compounds in such areas as Wana, Pakistan; and facilitating their movement to the target. Al Qa'ida leaders in Pakistan, such as Ayman al-Zawahiri, argued for the "need to concentrate on the method of martyrdom operations as the most successful way of inflicting damage against the opponent and the least costly to the Mujahedin in terms of casualties."

In addition, al Qa'ida played an important role in encouraging and implementing more-sophisticated IEDs, including remote-control detonators.[47] For example, there were a handful of al Qa'ida–run training facilities and IED factories in such places as Waziristan. They ranged from small facilities hidden within compounds that built IEDs to much larger "IED factories," which doubled as training centers and labs where recruits experimented with IED technology. One effective IED was the "TV bomb," which Iraqi groups first established. It was a shaped-charge mechanism that could be hidden under brush or debris on a roadside and set off by remote control from 300 yards or more. Taliban commanders received assistance from al Qa'ida and Iraqi groups on armor-penetrating weapons by disassembling rockets and rocket-propelled grenade rounds, removing the explosives and propellants, and repacking them with high-velocity shaped charges.[48] In addition, insurgent groups occasionally adopted brutal tactics, such as beheadings. In December 2005, for example, insurgents posted a video of the decapitation of an Afghan hostage on al Qa'ida–linked Web

---

[46] Hekmat Karzai and Seth G. Jones, "How to Curb Rising Suicide Terrorism in Afghanistan," *Christian Science Monitor*, July 18, 2006; Pamela Constable, "Gates Visits Kabul, Cites Rise in Cross-Border Attacks," *Washington Post*, January 17, 2007, p. A10; Jason Straziuso, "2007 Was a Year of Record Violence in Afghanistan, but U.S. Says Things Are Looking Up," Associated Press, January 1, 2008.

[47] Ali A. Jalali, "The Future of Afghanistan," *Parameters*, Vol. 36, No. 1, Spring 2006, pp. 4–19, p. 8.

[48] Interview with Afghan government officials, Kabul, Afghanistan, August 2006; Al Jazeera, interview with Mullah Dadullah, July 2005; Also see such press accounts as Sami Yousafzai and Ron Moreau, "Unholy Allies," *Newsweek*, September 26, 2005, pp. 40–42.

sites. This was the first-ever published video showing the beheading of an Afghan hostage.[49]

Al Qa'ida also played a role in connecting Afghan and Pakistani insurgent groups to the broader jihadi network, including in Iraq. With al Qa'ida's assistance, Islamic militants in Iraq provided information, through the Internet and face-to-face visits, on tactics to the Taliban and other insurgent groups. This included suicide tactics and various kinds of remote-controlled devices and timers. In addition, a small number of Pakistani and Afghan militants received military training in Iraq; Iraqi fighters met with Afghan and Pakistani extremists in Pakistan; and militants in Afghanistan increasingly used homemade bombs, suicide attacks, and other tactics honed in Iraq.

Conclusion

Despite initial success in capturing some al Qa'ida leaders, the United States failed to significantly weaken the organization. There was an increase in the number of attacks that involved al Qa'ida either directly or indirectly, an expansion of al Qa'ida's geographic reach, and an evolution of its organizational structure. Part of the reason was an overreliance on military force and the perception that there was a battlefield solution to a "war" on terror. But military force has rarely been effective against terrorist groups in the past. We now turn to what we believe is a more promising strategy.

---

[49] In what appeared to be a forced confession, Saeed Allah Khan stated, "I worked as a spy for the Americans along with four other people. The group received $45,000 and my share is $7,000" (Hekmat Karzai, *Afghanistan and the Globalisation of Terrorist Tactics*, Singapore: Institute of Defence and Strategic Studies, January 2006, p. 2).

CHAPTER SEVEN

Ending the "War" on Terrorism

The U.S. strategy after September 2001 was not effective in significantly weakening al Qa'ida by 2008. Some have argued that an effective strategy against al Qa'ida should include a broad range of tools that target the demand and supply side of the organization. As Rohan Gunaratna argued, for example, this strategy includes sanctions against state sponsors; the use of military and police forces against al Qa'ida's leaders, members, collaborators, and supporters; the resolution of regional conflicts in such locations as Kashmir and Palestinian territory; redressing grievances and meeting the legitimate aspirations of Muslims; and countering al Qa'ida's ideology.[1] Similarly, Daniel Byman noted that "there is no single strategy that can successfully defeat the jihadists. All heads of the hydra of terrorism must be attacked."[2]

A comprehensive strategy should indeed include a range of tools. The problem, however, is that a "kitchen-sink" approach does not prioritize a finite amount of resources and attention. Nor does it provide an assessment of what is most likely to be effective (and what is not). For example, economic sanctions are rarely effective in changing

---

[1] Gunaratna (2002, pp. 308–322); also see Rohan Gunaratna, "Combating the Al-Qaeda Associated Groups," in Doron Zimmermann and Andreas Wenger, eds., *How States Fight Terrorism: Policy Dynamics in the West*, Boulder, Colo.: Lynne Rienner Publishers, 2007, pp. 175–202.

[2] Daniel Byman, *The Five Front War: The Better Way to Fight Global Jihad*, Hoboken, N.J.: John Wiley and Sons, 2008, p. 44.

other states' behavior, including on issues related to terrorism.[3] In addition, the resolution of conflicts in such places as Kashmir and Palestinian territory may take generations and is not a primary reason for al Qa'ida's existence or support. We must therefore look elsewhere for an effective strategy that helps prioritize resources and attention. While numerous terrorist groups have ended because of a political solution, al Qa'ida's broad goals make this unlikely. Since its goal remains the establishment of a pan-Islamic caliphate, there is little reason to expect that a negotiated settlement with governments in the Middle East is possible.

Based on our analysis of how terrorist groups end, a more effective approach would be adopting a two-front strategy. First, policing and intelligence should be the backbone of U.S. efforts. In Europe, North America, North Africa, Asia, and the Middle East, al Qa'ida consists of an amorphous network of individuals who need to be tracked down and arrested. In Pakistan, for example, the most successful efforts to capture or kill al Qa'ida leaders after the September 2001 attacks—such as Khalid Sheikh Mohammed, Ramzi Binalshibh, Abu Faraj al-Libbi, and Abu Zubeida—occurred because of careful police and intelligence work, not military force. This strategy should include careful work abroad from such organizations as the CIA and FBI, as well as their cooperation with local police and intelligence agencies.

Second, military forces, but not necessarily U.S. military forces, are a necessary component when al Qa'ida is directly involved in an insurgency. Even in these cases, local military forces frequently have more legitimacy to operate than the United States does and a better understanding of the operating environment. This means a light U.S. footprint or none at all. The U.S. military can play a critical role in building indigenous capacity but should generally resist being drawn

---

[3] See, for example, Robert A. Pape "Why Economic Sanctions Do Not Work," *International Security*, Vol. 22, No. 2, Autumn 1997, pp. 90–136; Robert Al Pape, "Why Economic Sanctions Still Do Not Work," *International Security*, Vol. 23, No. 1, Summer 1998, pp. 66–77; T. Clifton Morgan and Valerie L. Schwebach, "Fools Suffer Gladly: The Use of Economic Sanctions in International Crises," *International Studies Quarterly*, Vol. 41, No. 1, March 1997, pp. 27–50; David Cortright and George A. Lopez, *The Sanctions Decade: Assessing UN Strategies in the 1990s*, Boulder, Colo.: Lynne Rienner Publishers, 2000.

into combat operations in Muslim countries, where its presence is likely to increase terrorist recruitment.

Changing the Symbols

The backbone of this two-front strategy should include focusing on careful police and intelligence work at home and abroad. This would include ending the notion of a "war" on terrorism and replacing it with such concepts as *counterterrorism*, which most governments with a significant terrorism problem use. This change might seem pedantic but would have significant symbolic importance. Moving away from military references would indicate that there was no battlefield solution to countering terrorism. Indeed, former secretary of defense Donald Rumsfeld had serious reservations about using the term *war on terror* because he was "concerned that the word 'war' led people to overemphasize the importance of the military instrument in this multidimensional conflict."[4] Individuals, such as Ayman al-Zawahiri and Osama bin Laden, should be viewed and described as criminals, not as holy warriors. "The notion of a war on terrorism," one British intelligence official told us, "suggests to Muslims abroad that the United States is fighting a war on Muslims. And the response has to be jihad, or holy war. War convinces people to do jihad."[5]

In Britain, for example, the government shunned the phrase *war on terror* despite a long history of dealing with such terrorist groups as the IRA. Hilary Benn, Britain's international-development secretary, argued that the phrase suggests that only military measures could be a useful response. "In the UK," he noted, "we do not use the phrase 'war on terror' because we can't win by military means alone and because this isn't one organized enemy with a clear identity and a coherent

---

4 Feith (2008, p. 87).

5 Interview with UK intelligence official by Seth G. Jones, Washington, D.C., January 23, 2008.

set of objectives."[6] The phrase raises public expectations—both in the United States and abroad—that there was a battlefield solution to the problem of terrorism. Similarly, the French government refused to refer to counterterrorism efforts as *war*, arguing that the term legitimized the terrorists.[7] And in Australia, government officials eschewed the use of the phrase *war on terror*.

This strategy is consistent with our findings. A transition to the political process is the most common reason that terrorist groups end (43 percent). The possibility of a political solution is linked to a key variable: the breadth of terrorist goals. Most terrorist groups that end because of politics seek narrow policy goals, such as policy and territorial change. If a transition to nonviolence is impossible or undesirable, policing is likely to be the most effective strategy to destroy terrorist groups (40 percent). The logic is that police and intelligence services have better training and information to penetrate and disrupt terrorist organizations. They are the primary arm of the government focused on internal security matters. Unlike the military, the police and intelligence agencies usually have a permanent presence in cities, towns, and villages; a better understanding of the threat environment in these areas; and better human intelligence.

Other strategies are less likely to be successful. The military is usually too blunt an instrument, and most soldiers are not trained to understand, penetrate, and destroy terrorist organizations; thus, 7 percent of terrorist groups have ended as a result of military action. In addition, 10 percent of the time, terrorist groups ended because their goals were achieved. When groups achieved victory, it was usually because they had narrow goals, such as policy or territorial change. No terrorist group that sought empire or social revolution ever achieved victory.

---

[6] Jane Perlez, "Briton Criticizes U.S.'s Use of 'War on Terror,'" *New York Times*, April 17, 2007, p. A10.

[7] Interview with members of l'Unité de Coordination de la Lutte Antiterroriste by Seth G. Jones, Monterey, Calif., January 22, 2008.

Policing and Intelligence

This strategy should include rebalancing U.S. resources and attention on police and intelligence work. It also means increasing budgets at the CIA, U.S. Department of Justice, and U.S. Department of State, and scaling back the U.S. Department of Defense's focus and resources on counterterrorism. U.S. special operations forces will remain critical, as will U.S. military operations to counter terrorist groups involved in insurgencies.

This also requires the development of a strategy with police and intelligence as its backbone. Unlike the military, local police and intelligence agencies usually have a permanent presence in cities and towns, a better understanding of local groups, and human sources. As Bruce Hoffman argued, a critical step in countering terrorist groups is for law-enforcement officials to

> develop strong confidence-building ties with the communities from which terrorists are most likely to come or hide in. . . . The most effective and useful intelligence comes from places where terrorists conceal themselves and seek to establish and hide their infrastructure.[8]

Some have argued that history has little to offer, since al Qa'ida's global breadth and decentralized organizational structure make it somewhat different from many other terrorist organizations, even religious ones. As Rohan Gunaratna argued, "Because there is no historical precedent for al Qaeda, the past offers very little guidance."[9] But this is not true. While al Qa'ida is different from many other terrorist organizations because of its global reach, its modus operandi is not atypical. Like other groups, its members need to communicate with each other, raise funds, build a support network, plan and execute attacks, and establish a base (or bases) of operation. Most of these nodes are vulnerable to penetration by police and intelligence agencies. The downside of this

---

[8] Hoffman (2006, p. 169).

[9] Gunaratna (2002, p. 296).

development is that eliminating key nodes in multiple places is more difficult than doing it one country.

Indeed, its organizational structure makes it vulnerable to a policing and intelligence strategy. This structure includes a bottom-up approach (encouraging independent thought and action from low-level operatives) and a top-down one (issuing orders and still coordinating a far-flung terrorist enterprise with both highly synchronized and autonomous moving parts). Al Qa'ida is a broad network. Successfully targeting this network requires a painstaking process of collecting intelligence on al Qa'ida, penetrating cells, and eventually arresting or killing its key members. As Mark Sageman argued, the most effective tools to defeating al Qa'ida and the global Salafi jihad "simply amount to good police work."[10] Unlike a hierarchical organization that can be eliminated through decapitation of its leadership, a network resists fragmentation because of its dense interconnectivity. A significant fraction of nodes can be randomly removed without much impact on its integrity. A network is vulnerable, however, at its hubs. If enough hubs are destroyed, the network breaks down into isolated, noncommunicating islands of nodes.[11] Hubs in a social network are vulnerable because most communications go through them. With good intelligence, law-enforcement authorities should be able to identify and arrest these hubs. This includes intercepting and monitoring terrorist communication through telephone, landline phone, email, facsimiles, and Internet chat rooms, as well as tracking couriers used by al Qa'ida officials in such places as Pakistan.

Police and intelligence services are best placed to implement these activities. This approach can include a range of steps: intelligence collection and analysis, capture of key leaders, and legal and other measures.

---

[10] Sageman (2004, p. 175).

[11] On networks and terrorism, see Seth G. Jones, "Fighting Networked Terror Groups: Lessons from Israel," *Studies in Conflict and Terrorism*, Vol. 30, No. 4, April 2007b, pp. 281–302; and Jackson (2006).

Intelligence Collection and Analysis

The first is intelligence collection and analysis. Intelligence is the principal source of information on terrorists. The police and intelligence agencies have a variety of ways of identifying terrorists, including signal intelligence (such as monitoring cell-phone calls) and human intelligence (such as using informants to penetrate cells). Human intelligence can provide some of the most useful actionable intelligence. But it requires painstaking work in recruiting informants who are already in terrorist organizations or placing informants not yet in them.

This means monitoring key individuals within al Qa'ida central, as well as key hubs in such places as Saudi Arabia, Iraq, Algeria, and the United Kingdom. The head of the UK security service, MI-5, noted in late 2007 that his organization identified at least 2,000 individuals "who we believed posed a direct threat to national security and public safety, because of their support for terrorism." Most were al Qa'ida inspired or assisted.[12] The greatest priority should be extensive penetration of terrorist networks. Recruitment of agents in place is sometimes difficult because of the strong emotional bonds among members of terrorist networks, making them reluctant to betray their friends and their faith. Local police and intelligence services are often better placed to recruit informants in al Qa'ida cells. In 2008, MI-5 estimated that it took roughly 18 months on average for an individual to become radicalized enough to conduct an attack. That period is fundamental for police and intelligence services to identify suspects, collect information, and arrest them.[13]

The best avenue for penetration often lies in recruiting from the pool of those who went through training but decided not to join the jihad or others who might associate with jihadists in such places as mosques. In the 2007 plot to target John F. Kennedy International Airport, for example, New York law-enforcement officials recruited an informant who attended the same Brooklyn mosque as Russell Defrei-

---

[12] Jonathan Evans, "Address to the Society of Editors by the Director General of the Security Service, Jonathan Evans," Society of Editors' "A Matter of Trust" conference, Radisson Edwardian Hotel, Manchester, UK, November 5, 2007.

[13] Interview with UK intelligence official by Seth G. Jones, January 23, 2008.

tas, one of the suspects. Defreitas said that he had met the informant at services at the mosque, took him into his confidence, and slowly disclosed his plan to attack Kennedy.[14] Imams of conservative or fundamentalist mosques who reject terrorism could be excellent sources of information on their congregations. They would be valuable allies to recruit, because they often know which members of their congregations are relatives or former friends of suspected terrorists. However, this radicalization does not always occur at mosques, since jihadists have become adept at evading detection. Mohammad Sidique Khan radicalized his group in the back of a van. Others may become radicalized in bookshops, on camping trips, or in other venues.

In Afghanistan, for example, police and intelligence officials focused on working with imams at mosques to counter Taliban and al Qa'ida recruitment efforts. As an Afghan intelligence report concluded, "The ease which the Taliban use the mullahs against us [needs to] be challenged." Consequently, the report concluded that

> this requires establishment of [a] relationship with every significant mullah in the country. . . . We should put our weight behind the nationalist ones and not allow the militant or fanatic ones to take over. This is only possible if we keep the nationalist ones on our pay-rolls.[15]

Since social bonds play a critical role in al Qa'ida's network, friends and relatives of identified terrorists need to be pursued and investigated wherever they reside. Especially important are those who were friends of a terrorist just before he or she started jihad, such as traveling to Pakistan for training. These friends may have helped transform him or her from an alienated Muslim into a dedicated terrorist. Arresting key individuals would degrade the network into isolated units or

---

[14] See *United States v Russell Defreitas, Kareem Ibrahim, Abdul Kadir, and Abdel Nur*, indictment, E.D.N.Y., filed June 28, 2007; U.S. Attorney's Office, Eastern District of New York, "Four Individuals Charged in Plot to Bomb John F. Kennedy International Airport," press release, Brooklyn, N.Y., June 2, 2007; and NYPD (2007).

[15] Amrullah Saleh, *Strategy of Insurgents and Terrorists in Afghanistan*, Kabul: National Directorate for Security, 2006, p. 16.

cliques. They would be less capable of mounting complex, large-scale operations, because they would lack expertise, logistical support, and financial support. Small-scale terrorist operations are difficult to end. But without spectacular successes to sustain their motivation, isolated operators would lose their enthusiasm. And this isolation would reduce terrorism to simple criminality. Winning the media war to label terrorists as criminals is especially important and virtually impossible to do in the face of a strategy based on military force.[16]

Working with local police and intelligence agencies is critical. They generally have better training and information to penetrate and disrupt terrorist organizations. They are the primary arm of the government focused on internal security matters.[17] Their mission should be to penetrate and seize terrorists and other criminals—their command structure, members, logistical support, and financial and political support—from the midst of the population. Local police and intelligence know the language, people, culture, and terrain better than U.S. agencies do. To paraphrase a U.S. special forces mantra, this strategy requires working "by, with, and through" local security forces.

Human intelligence is preferable to signal intelligence, since there are limitations to using technological means to monitor al Qa'ida movements. One good example is al Qa'ida's courier system. The organiztion adopted a four-tiered courier system to communicate among key members of the group and minimize detectability. Many al Qa'ida leaders have become more cautious in using cell phones, satellite phones, email, and other forms of communication that foreign intelligence services could easily track. The administrative courier network dealt with communication pertaining to the movement of al Qa'ida members' families and other administrative activities. The operational courier network dealt with operational instructions. Where possible, unwitting couriers were substituted for knowledgeable people to minimize detection. The media courier network was used for propaganda. Messages were sent in the form of CDs, videos, and leaflets to television networks, such as al Jazeera. Only al Qa'ida's top leadership, who usually did not pass writ-

---

[16] Sageman (2004, pp. 175–184).

[17] Trinquier (1964, p. 43); Galula (2005, p. 31).

ten messages to each other to maximize secrecy, used the final courier network. Normally, their most trusted couriers memorized messages and conveyed them verbatim.[18] The use of a sophisticated courier network places a premium on recruiting informants already in these organizations or placing informants in them.

For the United States, this approach requires providing foreign assistance to police and intelligence services abroad to improve their counterterrorism capacity. This means relying on the efforts of law enforcement and internal security forces of states where al Qa'ida is operating. The United States can help bolster the capabilities of foreign police and intelligence services abroad, as well as share intelligence information. Key locations where al Qa'ida has a foothold include Europe (such as Britain and the Netherlands), Algeria, Turkey, Saudi Arabia, Bangladesh, Indonesia, Pakistan, Afghanistan, and Iraq. The effort against al Qa'ida will hinge on the competence of local police and intelligence services in these countries to collect information, penetrate al Qa'ida cells, arrest or kill its members, and counter its propaganda machine. Working with locals is sometimes easier said than done, since not all states may cooperate. As Harvard law professor Philip Heymann pointed out, "Some states will lack the competence to really help, and states that do not believe in the cause will make efforts too half-hearted to be effective but real enough to be indistinguishable from sanctionable incompetence."[19] This is where other strategies, such as diplomacy and economic sanctions, can be useful in coercing states to support U.S. interests. In some cases, limited direct action may be inevitable.

Capture of Key Leaders

Next is the capture of key leaders and their support network. In democratic countries, this involves capturing key members and presenting the evidence in court. Terrorism involves the commission of violent crimes, such as murder and assault. The investigation, trial, and pun-

---

[18] Musharraf (2006, p. 221).

[19] Philip B. Heymann, "Dealing with Terrorism: An Overview," *International Security*, Vol. 26, No. 3, Winter 2001, pp. 24–38, p. 34.

ishment of perpetrators should be a matter for the wider criminal-justice system.[20] The barriers can sometimes be significant. Finding evidence that can be presented in court but that does not reveal sensitive information about sources and methods can be challenging. This is especially true if a terrorist has not perpetrated an attack yet. In many cases, it may be easier and more effective to arrest and punish terrorists for other offenses, such as drug trafficking, that have little direct connection to their terrorist activity. As one member of Antiterrorist Coordination Unit (l'Unité de Coordination de la Lutte Antiterroriste, or UCLAT) told us, "We have frequently detained possible terrorists for a number of crimes—such as criminal activity—that have little or nothing to do with terrorism. We can often build stronger legal cases against individuals by focusing on other crimes they have committed. The problem, of course, is that the punishment may be less severe."[21] In nondemocratic countries, the policing approach is often drastically different, because laws and norms of behavior may be different. Consequently, Pakistani and Saudi police and intelligence agencies have frequently used repressive measures to target al Qa'ida terrorists operating in their countries.[22]

The capture of terrorists—both low level and high level—is often a good source of information on leaders. Diaries, cell phones, and laptops can provide crucial information on code names of other terrorists, real names, addresses, phone numbers, and plans. For example, the 2004 capture of Muhammad Naeem Noor Khan led to a gold mine of information on al Qa'ida terrorist plots for the Pakistan government, United States, and other countries. He was a Pakistani national who was born in Karachi and earned a bachelor's degree in computer engineering in 2002. In March 2002, Khalid Sheikh Mohammed recruited

---

[20] See, for example, Clutterbuck (2004, pp. 142–144).

[21] Interview with member of l'Unité de Coordination de la Lutte Antiterroriste by Seth G. Jones, Monterey, Calif., June 27, 2007.

[22] On the dilemmas of U.S. assistance to the police of nondemocratic countries, see Seth G. Jones, Olga Oliker, Peter Chalk, C. Christine Fair, Rollie Lal, and James Dobbins, *Securing Tyrants or Fostering Reform? U.S. Internal Security Assistance to Repressive and Transitioning Regimes*, Santa Monica, Calif.: RAND Corporation, MG-550-OSI, 2006.

him. After two top al Qa'ida leaders—Ammar al-Baluchi and Tawfiq bin Attash—were captured, Khan became a key al Qa'ida official in Karachi. He was involved in training al Qa'ida operatives in Shakai, Pakistan. During this time, he remained closely associated with such al Qa'ida leaders as Abd al-Hadi al-Iraqi, Hamza al-Jawfi, Abu Faraj al-Libbi, and Abu Musab al-Baluchi. With help from the CIA, which had been tracking him, Pakistani intelligence arrested him on July 13, 2004. He—and his laptop—were a gold mine of information and gave Pakistan, the United States, and other countries vital information on al Qa'ida operations. The laptop contained the plans of Abu Issa al-Hindi (also known as Dhiren Barot), a senior member of al Qa'ida whom British authorities arrested for plotting attacks in the United States and UK. Under interrogation, Khalid Sheikh Mohammed acknowledged that he had told Khan to carry out reconnaissance of, and prepare a plan to attack, Heathrow Airport. After initial planning, Khan also suggested Canary Wharf and London's subway system as additional targets. Access to Khan's computer after his capture showed that their well-advanced plans included attacks on the headquarters of Citigroup and Prudential Insurance Group in New York, the United Nations headquarters in New York, and the International Monetary Fund and World Bank buildings in Washington.[23]

Legal and Other Measures

The third step is the development and passage of legal measures. This can involve criminalizing activities that are necessary for terrorist groups to function, such as raising money or recruiting members. It can also involve passing laws that make it easier for the intelligence and police services to conduct searches, engage in electronic surveillance, interrogate suspects, and monitor groups that pose a terrorist threat. It can include efforts to protect witnesses, juries, and judges from threats and intimidate. In democratic states, this inevitably leads to tension between civil liberties and security. As one scholar noted, "a democratic nation wants life, liberty, and unity as the products of its policies for dealing with terrorism, not just physical security. Focusing exclu-

[23] Musharraf (2006, pp. 241–242).

sively on a very popular desire for revenge . . . is likely to provide too little liberty and unity to be a sensible policy."[24] Since terrorist groups need to move money to multiple cells to help sustain their operations, attacking their finances or following financial leads once terrorists are captured has provided effective results. But there are challenges. The financial system known in the Islamic world as *hawala* exists outside the regulated international financial system. Individuals in Islamic communities around the world serve as go-betweens and facilitate the transfer of cash that is not taxed, recorded, or registered by banks. These informal *hawala* networks remain largely outside government control, and monitoring them presents a significant challenge to closing terrorist financial exchanges.[25]

Countering Ideology

Counterterrorism is just as much about hearts and minds as it is about policing and intelligence. It requires taking calculated actions that do not alienate Muslims. And it also requires effectively countering the ideology and messages of terrorist groups through what is often referred to as *information operations*. This includes the use of a variety of strategies and tools to counter, influence, or disrupt the message and operation of terrorist groups.[26] Local groups are almost always better placed to conduct information operations than the United States is. In addition to building local police and intelligence capacity, dealing with al Qa'ida also requires countering its ideological appeal. This includes countering the continued resonance of its message, its ability to attract recruits and replenish its ranks, and its capacity for continual regeneration and renewal. To do so, the United States needs to better understand the mind-set and minutia of the al Qa'ida movement, the ani-

---

[24] Philip B. Heymann, *Terrorism and America: A Commonsense Strategy for a Democratic Society*, Cambridge, Mass.: MIT Press, 1998, p. 153.

[25] On challenges in countering the *hawala* system, see Gary Berntsen, *First Directive*, unpublished manuscript, August 2007.

[26] On the use of *information operations* in a military context, see, for example, USJFCOM (2006).

mosity and arguments that underpin it, and indeed the regions of the world from which its struggle emanated.

Local groups are more likely to be effective in influencing locals and countering terrorist ideology than is the U.S. military or other international actors. It is critical to understand who holds power, whom the local population trusts, and where locals get their information—and then to target these forums. In some cases, such as in Afghanistan and Pakistan's tribal areas, religious leaders and tribal elders wield most of the power. This means providing assistance to credible indigenous groups, such as Muslim clerics or tribal elders, that can effectively counter jihadist propaganda. These groups do not necessarily have to be supportive of the United States, but they do need to oppose insurgents and have credible influence among the population. Much of this funding may have to be indirect and covert to protect their credibility. Assistance could be directed to indigenous media, political parties, student and youth organizations, labor unions, and religious figures and organizations that meet at least two criteria: (1) They have a notable support base in the local population, and (2) they oppose insurgent groups and insurgent ideology. This approach has some parallels with U.S. efforts during the Cold War to balance the Soviet Union by funding existing political, cultural, social, and media organizations in such areas as Central and Eastern Europe.[27]

In Afghanistan, for instance, mosques have historically served as a tipping point for major political upheavals. This led to a major effort by Afghan intelligence officials to focus on mosque leaders. As one Afghan intelligence report in 2006 concluded, "There are 107 mosques in the city of Kandahar out of which 11 are preaching anti-government themes. Our approach is to have all the pro-government mosques incorporated with the process and work on the eleven anti-government ones to change their attitude or else stop their propaganda and leave the area."[28] In addition, in July 2005, the Ulema Council of Afghani-

---

[27] See, for example, Angel Rabasa, Cheryl Benard, Lowell H. Schwartz, and Peter Sickle, *Building Moderate Muslim Networks,* Santa Monica, Calif.: RAND Corporation, MG-574-SRF, 2007.

[28] Saleh (2006, p. 8).

stan called on the Taliban to abandon violence and support the Afghan government in the name of Islam. They also called on the religious scholars of neighboring countries—including Pakistan—to help counter the activities and ideology of the Taliban and other insurgent organizations.[29] A number of Afghan Islamic clerics publicly supported the Afghan government and called the jihad un-Islamic.[30] Moreover, the Ulema Council and some Afghan ulama issued fatwas, or religious decrees, that unambiguously oppose suicide bombing. They argued that suicide bombing did not lead to an eternal life in paradise, did not permit martyrs to see the face of Allah, and did not allow martyrs to have the company of 72 maidens in paradise. These efforts were more effective than U.S.-led information operations, such as dropping leaflets.

This strategy was accomplished successfully during the Cold War when applied appropriately. During the Cold War, the United States believed that internal security assistance was critical to prevent certain countries from falling under Soviet influence.[31] The Office of Public Safety, which was established in 1962 in USAID, trained more than 1 million foreign police during its 13-year tenure.[32] President John F. Kennedy, for example, believed that Moscow sought to strengthen its international position by pursuing a strategy of subversion, indirect warfare, and agitation designed to install communist regimes in the developing world. In March 1961, President Kennedy told the U.S. Congress that the West was being "nibbled away at the periphery" by a Soviet strategy of "subversion, infiltration, intimidation, indirect or non-overt aggression, internal revolution, diplomatic blackmail, guerilla warfare or a series of limited wars."[33] He concluded that providing assistance to police and other internal security forces was critical

[29] "Religious Scholars Call on Taliban to Abandon Violence," *Pajhwok Afghan News*, July 28, 2005.

[30] "Taliban Claim Killing of Pro-Government Religious Scholars in Helmand," *Afghan Islamic Press*, July 13, 2005.

[31] Gaddis (1982); Brands (1993).

[32] Call (1998, p. 317); also see Huggins (1998, p. 111).

[33] Kennedy (1962).

to combat Soviet aggression, since they were the first line of defense against subversive forces. Robert Komer, President Kennedy's key National Security Council staff member on overseas internal security assistance, argued that viable foreign police in vulnerable countries were the necessary "preventive medicine" to thwart Soviet inroads.[34] Komer argued that the police were in regular contact with the population; could serve as an early warning against potential subversion; and could be used to control riots, demonstrations, and subversives before they became serious threats.

Military Force

Military force is sometimes necessary to end terrorist groups, especially when they are engaged in insurgencies. In most cases, however, local forces have been most effective to take the lead. Local forces—with assistance from intelligence units and special operations forces—can contest areas to regain government presence and control and then conduct military and civil-military programs to expand the control and edge out terrorists.[35] The focus should be on consolidating and holding ground that is clearly progovernment, deploying forces to conduct offensive operations, and holding territory once it is cleared. Holding territory has often been the most difficult facet of clear-and-hold strategies used against al Qa'ida and other groups in Afghanistan, Pakistan, and Iraq. Sufficient numbers of forces are needed to hold territory once it is cleared, or insurgents can retake it. Local forces may not always be government forces, as the United States discovered in Afghanistan

---

[34] William Rosenau, "The Kennedy Administration, U.S. Foreign Internal Security Assistance and the Challenge of 'Subterranean War,' 1961–63," *Small Wars and Insurgencies*, Vol. 14, No. 3, Autumn 2003, pp. 65–99, p. 80; also see Maxwell D. Taylor, "Address at International Police Academy Graduation," press release, U.S. Agency for International Development, December 17, 1965.

[35] The clear, hold, and build section draws extensively from U.S. Department of the Army and U.S. Marine Corps, *Counterinsurgency*, Washington, D.C., field manual 3-24, 2006; and Joseph D. Celeski, *Operationalizing COIN*, Hurlburt Field, Fla.: JSOU Press, report 05-2, 2005.

and Iraq. This suggests that the most vulnerable hubs of al Qa'ida may sometimes be local substate actors.

The data in Chapter Five indicated that, when insurgent groups have ended, they negotiated a settlement with the government nearly half the time. One-quarter of the time, the group achieved victory, and just under a quarter of the time, military forces defeated the insurgent group. A negotiated settlement with al Qa'ida is unlikely, since governments in North America, Europe, the Middle East, and Asia would never agree to this outcome. This means that, in cases in which al Qa'ida is involved in an insurgency, limited military force may be necessary. Force was necessary in Afghanistan in 2001, for example, to target al Qa'ida's base of operations. But U.S. military and intelligence forces acted primarily in support of the Northern Alliance, which conducted most of the ground fighting.[36] In the majority of cases, the United States should avoid direct, large-scale military force in the Muslim world to target al Qa'ida, or it risks increasing local resentment and creating new terrorist recruits.

The U.S. focus outside of its borders should be to work by, with, and through indigenous forces. As some jihadists have argued, direct military engagement with the United States has been good for the jihadi movement. It rallies the locals behind the movement and pits the fight between Islam and the West. One of al Qa'ida's primary objectives, then, is "to put America's armies, which occupy the region and set up military bases in it without resistance, in a state of war with the masses in the region. It is obvious at this very moment that it stirs up movements that increase the jihadi expansion and create legions among the youth who contemplate and plan for resistance."[37]

In addition, outside forces can rarely win insurgencies for local forces. First, outside military forces are unlikely to remain for the duration of any counterterrorist effort, at least as a major combatant force.

---

[36] On the overthrow of the Taliban regime, see Schroen (2005); Stephen D. Biddle, *Afghanistan and the Future of Warfare: Implications for Army and Defense Policy*, Carlisle, Pa.: Strategic Studies Institute, U.S. Army War College, November 2002; Gary Berntsen and Ralph Pezzullo, *Jawbreaker: The Attack on Bin Laden and Al Qaeda*, New York: Crown Publishers, 2005; and Bob Woodward, *Bush at War*, New York: Simon and Schuster, 2002.

[37] Naji (2006, p. 20).

This is especially true when terrorist groups are involved in an insurgency. Insurgencies are usually of short duration only if the indigenous government collapses at an early stage. An analysis of all insurgencies since 1945 shows that successful counterinsurgency campaigns last for an average of 14 years, and unsuccessful ones last for an average of 11 years. Many also end in a draw, with neither side winning. Insurgencies can also have long tails: Approximately 25 percent of insurgencies won by the government and 11 percent won by insurgents last more than 20 years.[38] Since indigenous forces eventually have to win the war on their own, they must develop the capacity to do so. If they do not develop this capacity, indigenous forces are likely to lose the war once international assistance ends.[39] Second, local forces usually know the population and terrain better than external actors and are better able to gather and exploit intelligence. Third, the population may interpret an outsider playing a lead role as an occupation, eliciting nationalist reactions that impede success.[40] Fourth, a lead indigenous role can provide a focus for national aspirations and show the population that they—and not foreign forces—control their destiny. Competent governments that can provide services to their population in a timely manner can best prevent and overcome terrorist groups.

---

[38] See David C. Gompert, John Gordon IV, Adam Grissom, David R. Frelinger, Seth G. Jones, Martin C. Libicki, Edward O'Connell, Brooke K. Stearns, and Robert E. Hunter, *War by Other Means: Building Complete and Balanced Capabilities for Counterinsurgency (RAND Counterinsurgency Study: Final Report)*, Santa Monica, Calif.: RAND Corporation, MG-595/2-OSD, 2008, Appendix A. On time, also see Galula (1964, p. 10).

[39] On rentier states, see Barnett R. Rubin, *The Fragmentation of Afghanistan: State Formation and Collapse in the International System*, 2nd ed., New Haven, Conn.: Yale University Press, 2002, pp. 81–105; Charles Tilly and Gabriel Ardant, *The Formation of National States in Western Europe*, Princeton, N.J.: Princeton University Press, 1975; and Hazem Beblawi and Giacomo Luciani, eds., *The Rentier State*, London and New York: Croom Helm, 1987.

[40] David Edelstein, "Occupational Hazards: Why Military Occupations Succeed or Fail," *International Security*, Vol. 29, No. 1, Summer 2004, pp. 49–91, p. 51.

Conclusion

The good news about countering al Qa'ida is that its probability of success in actually overthrowing any governments is close to zero. Al Qa'ida's objectives are virtually unachievable, and it has succeeded in alienating most governments in Asia, Europe, North America, South America, the Middle East, and Africa. Nor does it have a firm base of support, as do such groups with welfare services, such as Hizballah and Hamas. As al Qa'ida expert Peter Bergen concluded, "Making a world of enemies is never a winning strategy."[41]

But the bad news is that U.S. efforts against al Qa'ida have not been successful. Despite some successes against al Qa'ida, the United States has not significantly undermined its capabilities. Al Qa'ida has been involved in more attacks in a wider geographical area since September 11, 2001, including in such European capitals as London, than it was before that date. Its organizational structure has also evolved. This means that the U.S. strategy in dealing with al Qa'ida must change. A strategy based predominantly on military force has not been effective. Considering al Qa'ida's organizational structure and modus operandi, only a strategy based primarily on careful police and intelligence work is likely to be effective.

---

[41] Bergen (2008).

APPENDIX A

End-of-Terror Data Set

Table A.1 provides information on the organizations included in our data set.

Table A.1
Organizations in the Data Set

Organization	Operating	Peak Size	Econ.[a]	Regime[b]	Type[c]	Goal[d]	Ended[e]
1920 Revolution Brigades (Iraq)	2003–	10s	LM	NF	R	RC	—
23rd of September Communist League (Mexico)	1973–1982	100s	UM	PF	LW	RC	PO
28 May Armenian Organization (Turkey)	1977	10s	LM	F	N	TC	S
2nd of June Movement (Federal Republic of Germany)	1975–1981	10s	H	F	LW	PC	S
Abdurajak Janjalani Brigade (Philippines)	1999	10s	LM	F	R	PC	S
Abu al-Abbas (Iraq)	2004	10s	LM	NF	R	PC	S

[a] Designations correspond to the World Bank analytical classification for each terrorist group's base country of operation in the year in which the use of terror ended. Current World Bank classifications have been employed for terrorist groups deemed active at the end of 2006. In 2006, the World Bank used the following thresholds for classifying national economies (in 2005 gross national income [GNI] per capita, calculated using the World Bank Atlas Method): L = low income ($875 or less); LM = lower middle income ($876–$3,465); UM = upper middle income ($3,466–$10,725); and H = high income ($10,726 or more).

[b] Designations correspond to the Freedom House classification of each terrorist group's base country of operation in the year in which the use of terror ended. Current Freedom House classifications have been employed for terrorist groups deemed active at the end of 2006. The status designations of F (free), PF (partly free), and NF (not free) are determined by the combination of the political-rights and civil-liberties ratings for each country or territory. These categories contain numerical ratings between 1 and 7 for each country or territory, with 1 representing the freest and 7 the least free. The aggregate score of 2 to 14 indicates the general state of freedom in each country or territory.

[c] R = religious. LW = left-wing. N = nationalist. RW = right-wing.

[d] RC = regime change. TC = territorial change. PC = policy change. E = empire. SR = social revolution. SQ = status quo.

[e] PO = policing. S = splintering. PT = politics. V = victory. MF = military force.

Table A.1—Continued

Organization	Operating	Peak Size	Econ.[a]	Regime[b]	Type[c]	Goal[d]	Ended[e]
Abu al-Rish Brigades (West Bank/Gaza)	1993–	100s	LM	PF	N	TC	—
Abu Bakr al-Siddiq Fundamentalist Brigades (Iraq)	2004	10s	LM	NF	R	PC	S
Abu Hafs al-Masri Brigade (Spain, UK)	2003–	10s	H	F	R	E	—
Abu Nayaf al-Afghani (Spain)	2004	10s	H	F	R	E	S
Abu Nidal Organization (Iraq, Libya, Syria)	1974–2002	100s	LM	NF	N	TC	PO
Abu Sayyaf Group (Philippines)	1991–	100s	LM	PF	R	E	—
Accolta Nazinuale Corsa (France)	2002–2003	10s	H	F	N	TC	S
Achik National Volunteer Council (Bangladesh, Burma, India)	1995–	100s	L	F	N	TC	—
Actiefront Nationistisch Nederland (Netherlands)	1992	10s	H	F	RW	PC	PT
Action Commitee of WInegrowers (France)	1999–	10s	H	F	LW	PC	—
Action Directe (France)	1979–1987	100s	H	F	LW	SR	PO
Action Front for the Liberation of the Baltic Countries (France)	1977	10s	H	F	N	SQ	V
Action Group Extreme Beate (Denmark)	2005	10s	H	F	LW	PC	PO
Aden Abyan Islamic Army (Yemen)	1994–	10s	L	PF	R	RC	—
Adivasi Cobra Force (India)	1996–	100s	L	F	N	PC	—
Affiche Rouge (France)	1981–1986	10s	H	F	LW	SR	PO

Table A.1—Continued

Organization	Operating	Peak Size	Econ.[a]	Regime[b]	Type[c]	Goal[d]	Ended[e]
African National Congress (S. Africa)	1961–1990	10,000s	UM	PF	N	PC	V
Akhil Krantikari (Nepal)	1995–	10,000s	L	PF	LW	RC	—
al-Arifeen (Pakistan)	2002–	100s	L	NF	R	TC	—
al-Ahwal Brigades (Iraq)	2005–	10s	LM	NF	R	RC	—
al-Aqsa Martyrs Brigades (Israel, West Bank/Gaza)	2000–	100s	LM	PF	N	TC	—
al-Badr (the first) (Pakistan)	1971	100s	L	PF	N	SQ	MF
al-Badr (the second) (Pakistan)	1998–	100s	L	NF	R	TC	—
Albanian National Army (Macedonia)	2002–	10s	LM	PF	N	TC	—
al-Bara bin Malek Brigades (Iraq, Jordan)	2005–	100s	LM	NF	R	RC	—
al-Barq (India, Pakistan)	1978–2002	100s	L	NF	N	TC	S
al-Borkan Liberation Organization (Italy)	1984–1985	10s	H	F	N	RC	S
Alejo Calatayu (Bolivia)	1987	10s	LM	F	LW	PC	PT
Alex Boncayao Brigade (Philippines)	1984–	100s	LM	PF	LW	RC	—
al-Faran (India)	1995	10s	L	PF	N	TC	S
al-Faruq Brigades (Iraq)	2003–	10s	LM	NF	R	RC	—
al-Fatah (Israel, West Bank/Gaza)	1958–	10,000s	LM	PF	N	TC	—

Table A.1—Continued

Organization	Operating	Peak Size	Econ.[a]	Regime[b]	Type[c]	Goal[d]	Ended[e]
al-Fuqra (Canada, Pakistan, U.S.)	1980–	1,000s	H	F	R	SR	—
al-Gama'a al-Islamiyya (Egypt, Afghanistan)	1977–	100s	LM	NF	R	E	—
al-Hadid (India)	1994	10s	L	F	N	TC	S
al-Haramayn Brigades (Saudia Arabia)	2003–	10s	H	NF	R	RC	—
Ali bin Abu Talib Jihad Organization (Iraq)	2005–	10s	LM	NF	R	RC	—
Alianca Libertadora Nacional (Brazil)	1968–1970	100s	LM	PF	LW	RC	PO
al-Intiqami al-Pakistani (Pakistan)	2002	10s	L	NF	R	E	PO
al-Ittihaad al-Islami (Ethiopia, Kenya, Somalia)	1989–1996	1,000s	L	NF	R	E	MF
All Burma Students' Democratic Front (Burma, Thailand)	1988–	100s	L	NF	LW	RC	—
All Tripura Tiger Force (Bangladesh, India)	1990–	100s	L	F	N	TC	—
al-Madina (India)	2002–	10s	L	F	N	TC	—
al-Mansoorain (India, Pakistan)	2003–	100s	L	F	N	TC	—
al-Nawaz (Pakistan)	1999–2000	10s	L	NF	N	RC	PO
al Qa'ida (Pakistan, Afghanistan, and others)	1988–	1,000s	L	NF	R	E	—
al Qa'ida in the Arabian Peninsula (Saudi Arabia)	2004–	100s	H	NF	R	E	—

Table A.1—Continued

Organization	Operating	Peak Size	Econ.[a]	Regime[b]	Type[c]	Goal[d]	Ended[e]
al Qa'ida Organization in the Land of the Two Rivers (Iraq, Jordan)	2004–	1,000s	LM	NF	R	RC	—
al-Qanoon (Pakistan)	2002	100s	L	NF	R	RC	S
al-Quds Brigades (Syria, Israel, West Bank/Gaza)	1978–	100s	LM	PF	R	TC	—
al-Sadr Brigades (Syria, Lebanon)	1978–	10s	UM	PF	R	PC	—
al-Saiqa (Syria, Israel, West Bank/Gaza)	1966–	100s	L	NF	N	TC	—
al-Umar Mujahideen (Pakistan)	1989–	100s	L	NF	R	TC	—
al-Zulfikar (Afghanistan, Libya, Pakistan, India, Syria)	1977–1981	10s	L	NF	N	RC	PO
Amal (Lebanon)	1975–	100s	UM	PF	N	RC	—
American Front (U.S.)	1990–1995	10s	H	F	RW	SR	PO
Ananda Marga (Australia, India)	1955–1979	100s	L	F	R	SR	PT
Anarchist Faction (Greece)	1999–2000	10s	H	F	LW	SR	S
Anarchists (Spain)	2000	10s	H	F	LW	SR	PT
Andres Castro United Front (Nicaragua)	1995–2002	100s	L	PF	LW	PC	PO
Angry Brigade (Italy)	1999	10s	H	F	LW	SR	PT
Animal Liberation Front (Canada, U.S., UK)	1976–	10s	H	F	LW	PC	—

Table A.1—Continued

Organization	Operating	Peak Size	Econ.[a]	Regime[b]	Type[c]	Goal[d]	Ended[e]
Ansar al-Islam (Iraq)	2001–	100s	LM	NF	R	RC	—
Ansar al-Jihad (Iraq)	2004–	10s	LM	NF	R	RC	—
Ansar Allah (Lebanon)	1994	100s	LM	PF	R	RC	S
Ansar al-Sunnah Army (Iraq)	2003–	100s	LM	NF	R	RC	—
Anti-American Arab Liberation Front (Federal Republic of Germany)	1986	10s	H	F	N	PC	PT
Anti-Armenian Organization (France)	1984	10s	H	F	N	PC	PT
Anticapitalist Attack Nuclei (Italy)	2001	10s	H	F	LW	SR	S
Anti-Communist Command (Indonesia)	2000–	10s	LM	F	RW	PC	—
Anti-Imperialist Cell (Germany)	1994–1996	10s	H	F	LW	SR	PO
Anti-Imperialist Group Liberty for Mumia Abu Jamal (Germany)	1995	10s	H	F	LW	PC	PT
Anti-Imperialist International Brigade (Lebanon)	1986–1988	10s	LM	NF	LW	SR	PO
Anti-Imperialist Patrols for Proletariat Internationalism (Italy)	1983	10s	H	F	LW	RC	S
Anti-Imperialist Territorial Nuclei for the Construction of the Fighting Communist Party (Italy)	1995–	10s	H	F	LW	RC	—
Anti-Power Struggle (Greece)	1995–2001	10s	H	F	LW	SR	S

Table A.1—Continued

Organization	Operating	Peak Size	Econ.[a]	Regime[b]	Type[c]	Goal[d]	Ended[e]
Anti-Racist Guerrilla Nuclei (Italy)	1999	10s	H	F	LW	RC	S
Anti-Terrorist Liberation Group (France, Spain)	1983–1987	100s	H	F	RW	SQ	PO
Anti-Zionist Movement (Italy)	1999–2000	10s	H	F	RW	PC	PT
Apo's Revenge Hawks (Turkey)	1999	10s	LM	PF	N	TC	S
Apo's Youth Revenge Brigades (Turkey)	1999	10s	LM	PF	N	TC	S
April 19 Movement (Colombia)	1970–1990	1,000s	LM	PF	LW	RC	PT
Arab Communist Organization (Lebanon, Syria)	1974–1977	10s	LM	PF	LW	PC	PO
Arab Communist Revolutionary Party (Jordan)	1990–1991	10s	LM	PF	LW	PC	S
Arab Fedayeen Cells (Lebanon)	1986	10s	LM	PF	N	TC	S
Arab Liberation Front (Iraq, Israel, West Bank/Gaza)	1969–1986	100s	UM	NF	N	TC	S
Arab Nationalist Youth for the Liberation of Palestine (Libya)	1974	10s	H	NF	N	TC	PT
Arab Revenge Organization (Lebanon)	1984	10s	LM	PF	N	PC	S
Arab Revolutionary Front (France)	1986	10s	H	F	N	PC	S
Arabian Peninsula Freemen (Belgium)	1989	10s	H	F	N	RC	S
Arbav Martyrs of Khuzestan (Iran)	2005	10s	LM	NF	N	TC	S
Argentine Anti-Communist Alliance (Argentina)	1974–1976	100s	UM	NF	RW	SQ	PT

Table A.1—Continued

Organization	Operating	Peak Size	Econ.[a]	Regime[b]	Type[c]	Goal[d]	Ended[e]
Arizona Patriots (U.S.)	1984–1986	10s	H	F	RW	RC	PO
Armata Corsa (France)	1999–	10s	H	F	N	TC	—
Armata di Liberzione Naziunale (France)	2002–2003	10s	H	F	N	TC	V
Armed Commandos of Liberation (Puerto Rico)	1968–1972	10s	UM	F	N	RC	S
Armed Communist League (Mexico)	1972	10s	LM	PF	LW	RC	S
Armed Forces of National Liberation (Puerto Rico, U.S.)	1974–1985	10s	H	F	N	RC	PO
Armed Forces of National Resistance (El Salvador)	1975–1991	100s	LM	PF	LW	PC	PT
Armed Islamic Group (Algeria)	1992–2005	1,000s	LM	NF	R	RC	MF
Armed Revolutionary Action (Mozambique, Portugal)	1960–1974	100s	UM	PF	N	RC	V
Armed Revolutionary Left (Ecuador)	2004–	10s	LM	PF	LW	PC	—
Armenian Red Army (Lebanon, Netherlands)	1982	10s	LM	PF	N	TC	PO
Armenian Resistance Group (Belgium, Netherlands, Armenia)	1995	10s	H	F	N	TC	V
Armenian Revolutionary Army (Belgium, Portugal, U.S., Austria, Canada, Turkey, Armenia)	1970–1985	100s	H	F	N	TC	S
Armenian Secret Army for the Liberation of Armenia (Armenia, Lebanon)	1980–1997	100s	UM	NF	N	TC	V

Table A.1—Continued

Organization	Operating	Peak Size	Econ.[a]	Regime[b]	Type[c]	Goal[d]	Ended[e]
Army for the Liberation of Rwanda (Democratic Republic of the Congo)	1994–2002	1,000s	L	NF	N	RC	S
Army of God (U.S.)	1982–	10s	H	F	R	PC	—
Army of the Corsican People (France)	2004–	10S	H	F	N	TC	—
Army of the Followers of Sunni Islam (Iraq)	2004–	10s	LM	NF	R	RC	—
Arnoldo Camu Command (Chile)	1989	10s	LM	PF	LW	PC	PT
Aryan Nations (U.S.)	1979–2004	100s	H	F	RW	SR	PO
Asbat al-Ansar (Lebanon)	1989–	100s	UM	PF	R	RC	—
Aum Shinrikyo (Australia, Germany, Indonesia, Japan, Russia, Taiwan, U.S.)	1984–2000	10,000s	H	F	R	SR	PO
Autonomia Sinistra Ante Parlamentare (Italy)	1989	10s	H	F	LW	RC	S
Autonomous Cells (Federal Republic of Germany)	1987	10s	H	F	LW	SR	PT
Autonomous Decorators (Germany)	1999	10s	H	F	LW	SR	PT
Autonomous Intervention Collective Against the Zionist Presence in France and Against the Israel-Egyptian Peace Treaty (France)	1979	10s	H	F	N	TC	S
Azad Hind Sena (India)	1982	10s	L	F	N	PC	PO
Baader-Meinhof Group (Federal Republic of Germany)	1968–1977	100s	H	F	LW	SR	PO

Table A.1—Continued

Organization	Operating	Peak Size	Econ.[a]	Regime[b]	Type[c]	Goal[d]	Ended[e]
Babbar Khalsa International (India)	1978–	100s	L	F	N	TC	—
Badr Forces (West Bank/Gaza)	2001	10s	LM	PF	N	TC	S
Bagramyan Battalion (Georgia)	1998	10s	LM	PF	N	TC	PT
Baloch Liberation Army (Pakistan)	2000–	1,000s	L	NF	N	TC	—
Basque Fatherland and Freedom (Spain)	1958–	1,000s	H	F	N	TC	—
Battalion of the Martyr Abdullah Azzam (Egypt, Jordan, Syria)	2004–	100s	LM	NF	R	E	—
Bersatu (Thailand)	1989–	10s	LM	NF	N	TC	—
Birsa Commando Force (India)	1996–2004	10s	L	F	N	PC	PT
Black December (Pakistan)	1973	10s	L	PF	N	PC	PT
Black Friday (Turkey)	1988	10s	LM	PF	N	PC	PT
Black Hand (Lebanon)	1983	10s	LM	PF	R	RC	S
Black Liberation Army (U.S.)	1971–1985	100s	H	F	LW	SR	PO
Black Panthers (U.S.)	1966–1972	1,000s	H	F	LW	SR	PO
Black Panthers (West Bank/Gaza)	2005–	10s	LM	PF	N	TC	—
Black September (Jordan, Lebanon, West Bank/Gaza)	1971–1974	100s	LM	NF	N	TC	S
Black Star (Greece)	1999–2002	10s	H	F	LW	SR	S

Table A.1—Continued

Organization	Operating	Peak Size	Econ.[a]	Regime[b]	Type[c]	Goal[d]	Ended[e]
Black Widows (Russia)	2000–	10s	UM	BF	N	TC	—
Bodo Liberation Tigers (Bhutan, India)	1996–2003	1,000s	L	F	N	TC	V
Boere Aanvals Troepe (S. Africa)	1996–1997	10s	UM	F	RW	TC	PO
Bolivarian Guerilla Movement (Venezuela)	2003	10s	UM	PF	LW	SR	S
Bolivarian Liberation Forces (the first) (Venezuela)	1992	10s	UM	PF	LW	RC	PT
Bolivarian Liberation Forces (the second) (Venezuela)	2002–	100s	UM	PF	LW	SR	—
Borok National Council of Tripura (Bangladesh, India)	2000–	100s	L	F	N	TC	—
Breton Revolutionary Army (France)	1971–2000	10s	H	F	N	TC	PO
Brigade 313 (Pakistan)	2003–	10s	L	NF	R	RC	—
Brigades of Imam al-Hassan al-Basri (Iraq)	2005–	10s	LM	NF	Reigious	RC	—
Brigades of Martyr Ahmed Yassin (Iraq)	2004	10s	LM	NF	N	RC	S
Brigades of the Victorious Lion of God (Iraq)	2004–	10s	LM	NF	R	RC	—
Brother Julian (Bolivia)	1987	10s	LM	F	LW	PC	PT
Cambodian Freedom Fighters (Cambodia, U.S.)	1998–2001	10s	L	NF	LW	RC	PO
Carapaica Revolutionary Movement (Venezuela)	2002–	10s	UM	PF	LW	SQ	—
Catholic Reaction Force (Northern Ireland, UK)	1983–	10s	H	F	N	TC	—

Table A.1—Continued

Organization	Operating	Peak Size	Econ.[a]	Regime[b]	Type[c]	Goal[d]	Ended[e]
Cell for Internationalism (Austria)	1995	10s	H	F	LW	RC	S
Chadian People's Revolutionary Movement (Chad, Congo)	1982–1988	1,000s	L	NF	N	RC	PT
Charles Martel Group (France)	1975–1983	100s	H	F	RW	PC	PT
Che Guevara Anti-Imperialist Command (Argentina)	2005	10s	UM	F	LW	SR	PT
Che Guevara Brigade (Argentina)	1976–1990	10s	LM	F	LW	PC	S
Chilean Committee of Support for the Peruvian Revolution (Chile)	1992	10s	LM	F	LW	SR	S
Chukakuha (Japan)	1963–	100s	H	F	LW	PC	—
Cinchoneros Popular Liberation Movement (Honduras, Nicaragua)	1980–1991	100s	L	F	LW	PC	MF
Clandestini (France)	1999	10s	H	F	N	TC	S
Clandestini Corsi (France)	1999–	100s	H	F	RW	PC	—
Coalition of National Brigades (Haiti)	1973	10s	L	NF	LW	RC	PT
Colonel Karuna Faction (Sri Lanka)	2004–	1,000s	LM	PF	N	TC	—
Comando Internacionalista Simon Bolivar (Mexico)	1986–1987	10s	LM	PF	LW	PC	PT
Comando Jaramillista Morelense 23 de Mayo (Mexico)	2004	10s	UM	F	LW	PC	PT
Combatant Proletarian Nucleus (Italy)	2002	10s	H	F	LW	RC	S

Table A.1—Continued

Organization	Operating	Peak Size	Econ.[a]	Regime[b]	Type[c]	Goal[d]	Ended[e]
Comité Argentino de Lucha Anti-Imperialisto (Argentina)	1971–1972	10s	UM	PF	LW	SR	PT
Committee for Liquidation of Computers (France)	1983	10s	H	F	LW	SR	S
Committee for the Security of the Highways (Israel, West Bank/Gaza)	1998–2001	10s	LM	PF	N	TC	PT
Committee of Coordination (France)	1972	10s	H	F	N	PC	PT
Committee of Solidarity with Arab and Middle East Political Prisoners (France, Iran, Lebanon)	1985–1986	100s	H	F	N	PC	PO
Communist Combatant Cells (Belgium)	1985	10s	H	F	LW	SR	PO
Communist Party of India-Maoist (India)	2004–	1,000s	L	F	LW	TC	—
Communist Party of Nepal-Maoist (India, Nepal)	1996–2006	10,000s	L	PF	LW	RC	PT
Communist Revoluitionaries in Europe (Netherlands)	1999	10s	H	F	N	TC	PT
Communist Workers Movement (Turkey)	2001–2003	10s	LM	PF	LW	RC	PO
Conscientious Arsonists (Greece)	1997–1998	10s	H	F	LW	SR	S
Continuity Irish Republican Army (Ireland, Northern Ireland, UK)	1986–	100s	H	F	N	TC	—
Cooperative of Hand-Made Fire and Related Items (Italy)	2001–	10s	H	F	LW	SR	—
Corsican Revolutionary Armed Forces (France)	1992–	10s	H	F	N	TC	—

Table A.1—Continued

Organization	Operating	Peak Size	Econ.[a]	Regime[b]	Type[c]	Goal[d]	Ended[e]
Counter-Revolutionary Solidarity (Guatemala)	1983	10s	LM	NF	RW	SQ	PT
Covenant, Sword, and Arm of the Lord (U.S.)	1978–1985	10s	H	F	RW	SR	PO
Croatian Freedom Fighters (U.S.)	1976–1982	10s	H	F	N	TC	PO
Cypriot Nationalist Organization (Cyprus)	2004	10s	H	F	N	SQ	PT
Dagestan Liberation Army (Russia)	1999–2004	10s	UM	NF	N	TC	PO
Dagestani Shari'ah Jamaat (Russia)	2002–	100s	UM	NF	N	TC	—
Dario Santillan Command (Argentina)	2004	10s	UM	F	LW	PC	PT
de Fes (France, Morocco)	1994	10s	LM	PF	R	RC	PO
December 20 Movement (Panama)	1990–1992	100s	LM	PF	RW	RC	PT
Democratic Front for the Liberation of Palestine (Israel, West Bank/Gaza)	1969–	100s	LM	PF	N	TC	—
DHKP/C (Turkey)	1994–	100s	UM	PF	LW	RC	—
Dima Halam Daoga (India)	1996–	100s	L	F	N	TC	—
Divine Wrath Brigades (Iraq)	2004–	10s	LM	NF	R	RC	—
Dukhtaran-e-Millat (India)	1987–	10s	L	F	R	TC	—
Earth Liberation Front (Canada, U.S., UK)	1992–	10s	H	F	LW	PC	—

Table A.1—Continued

Organization	Operating	Peak Size	Econ.[a]	Regime[b]	Type[c]	Goal[d]	Ended[e]
East Turkistan Liberation Organization (China, Kyrgyzstan)	2002–	100s	LM	NF	N	TC	—
Eastern Turkistan Islamic Movement (China)	1990–	100s	LM	NF	N	TC	—
Ecuadorian Rebel Force (Ecuador)	2001	10s	LM	PF	R	PC	—
Egyptian Islamic Jihad (Egypt, Afghanistan)	1978–	100s	LM	NF	R	RC	—
Egypt's Revolution (Egypt)	1984–1989	10s	LM	PF	N	PC	PO
Ejercito del Pueblo en Armas (Venezuela)	2002–	10s	UM	PF	LW	SQ	—
Eritrean Islamic Jihad Movement (Eritrea, Ethiopia, Sudan)	1980–	1,000s	L	NF	R	RC	—
Eritrean Liberation Front (Eritrea, Ethiopia)	1960–1991	10,000s	L	PF	N	RC	V
Eritrean People's Liberation Front (Eritrea, Ethiopia)	1970–1991	10,000s	L	PF	N	RC	V
Ethiopian People's Revolutionary Army (Ethiopia, Sudan)	1976–1988	1,000s	L	NF	LW	RC	PT
Ethnocacerista (Peru)	2000–	100s	LM	F	RW	RC	—
Eva Peron Organization (Argentina)	1990	10s	LM	F	LW	PC	PT
Evan Mecham Eco-Terrorist International Conspiracy (U.S.)	1986–1989	10s	H	F	LW	PC	PO
Extraditables, The (Colombia)	1987–1991	10s	LM	PF	LW	PC	V
EZLN (Mexico)	1983–2005	1,000s	UM	F	LW	PC	PT
Farabundo Marti National Liberation Front (El Salvador)	1979–1991	10,000s	LM	PF	LW	RC	PT

Table A.1—Continued

Organization	Operating	Peak Size	Econ.[a]	Regime[b]	Type[c]	Goal[d]	Ended[e]
Fatherland and Liberty Nationalist Front (Chile)	1999–2000	10s	UM	F	RW	PC	V
February 28 Popular Leagues (El Salvador)	1978–1991	100s	LM	PF	LW	RC	PT
Fedayeen Khalq (Iran)	1979–1988	10s	UM	NF	LW	RC	PO
Fighting Jewish Organization (Israel, West Bank/Gaza)	1993–1995	10s	H	F	N	TC	S
First of October Antifascist Resistance Group (Spain)	1975–2006	100s	H	F	LW	RC	PO
Five Cs (Italy)	2002–	10s	H	F	LW	SR	—
For a Revolutionary Perspective (Switzerland)	2001	10s	H	F	LW	SR	PT
Francs Tireurs (France)	1991–1998	10s	H	F	LW	PC	PO
Free Aceh Movement (Indonesia, Malaysia, Sweden)	1975–2005	1,000s	LM	PF	N	TC	PT
Free Greeks (Greece)	1967–1974	100s	UM	F	N	RC	V
Free Papua Movement (Indonesia, Papua New Guinea)	1963–	100s	LM	PF	N	TC	—
Free People of Galilee (Israel)	2003	10s	H	F	N	TC	PO
Free South Moluccan Youth's (Netherlands)	1975–1978	10s	H	F	N	TC	PO
Free Vietnam Revolutionary Group (Philippines, Thailand, Vietnam, U.S.)	2001	10s	LM	F	N	RC	S
Freedom for Mumia Abu-Jamal (Switzerland)	1999	10s	H	F	LW	PC	PT

Table A.1—Continued

Organization	Operating	Peak Size	Econ.[a]	Regime[b]	Type[c]	Goal[d]	Ended[e]
Frente de Liberacion Nacional del Vietnam del Sur (Argentina)	1968	10s	UM	PF	LW	PC	PT
Front for Defenders of Islam (Indonesia)	1997–	1,000s	LM	F	R	SR	—
Front for the Liberation of the Cabinda Enclave (Angola)	1963–	100s	LM	NF	N	TC	—
Front for the Liberation of Lebanon from Foreigners (Lebanon)	1977–1983	100s	LM	PF	N	PC	S
Front for the Liberation of the French Somali Coast (Djibouti, Somalia)	1967–1977	1,000s	L	PF	N	TC	V
Fronte di Liberazione Naziunale di a Corsica (France)	1976–	100s	H	F	N	TC	—
Gazteriak (France)	1994–2000	1,000s	H	F	N	TC	S
Generation of Arab Fury (Iran, Kuwait)	1989	10s	UM	NF	R	RC	PO
Global Intifada (Sweden)	2002–	10s	H	F	LW	SR	—
God's Army (Burma, Thailand)	1997–2001	100s	L	NF	N	TC	MF
Gracchus Babeuf (France)	1990–1991	10s	H	F	LW	PC	PT
Greek Bulgarian Armenian Front (Australia)	1986	10s	H	F	N	TC	PO
Group Bakunin Gdansk Paris Guatemala Salvador (France)	1981–1982	100s	H	F	LW	SR	S
Group of Guerilla Compbatants of Jose Maria Morelos y Pavon (Mexico)	2001–	10s	UM	F	LW	RC	—
Group of Popular Combatants (Ecuador)	1994–	10s	LM	PF	LW	PC	—

Table A.1—Continued

Organization	Operating	Peak Size	Econ.[a]	Regime[b]	Type[c]	Goal[d]	Ended[e]
Group of the Martyrs Mostafa Sadeki and Ali Zadeh (Iran, Switzerland)	1993	10s	LM	NF	N	PC	PT
Guadeloupe Liberation Army (Guadeloupe)	1980–1981	10s	LM	F	N	RC	PT
Guardsmen of Islam (Germany, Iran)	1980–1984	10s	UM	PF	R	PC	PO
Guatemalan Labor Party (Guatemala)	1952–1996	1,000s	LM	PF	LW	RC	PT
Guatemalan National Revolutionary Unity (Guatemala)	1982–1996	10,000s	LM	PF	LW	RC	PT
Guerrilla Army of the Poor (Guatemala)	1972–1996	1,000s	LM	PF	LW	RC	PT
Guevarista Revolutionary Army (Colombia)	1993–	100s	LM	PF	LW	RC	—
Hamas (Israel, West Bank/Gaza)	1987–	1,000s	LM	PF	N	TC	—
Hammerskin Nation (U.S.)	1987–	100s	H	F	RW	SR	—
Harakat al-Shuhada'a al-Islamiyah (Libya)	1996–	100s	UM	NF	R	RC	—
Harakat ul-Jihad-i-Islami (Pakistan)	1980–	100s	L	NF	R	TC	—
Harakat ul-Jihad-i-Islami/Bangladesh (Bangladesh)	1992–	1,000s	L	PF	R	RC	—
Harakat ul-Mujahidin (Pakistan)	1985–	100s	L	NF	R	TC	—
Harakat ul-Ansar (India, Pakistan)	1993–2002	100s	L	F	N	TC	S
Hector Riobe Brigade (Haiti, U.S.)	1982–1984	10s	H	F	LW	RC	PO

Table A.1—Continued

Organization	Operating	Peak Size	Econ.[a]	Regime[b]	Type[c]	Goal[d]	Ended[e]
Heroes of Palestine (Colombia)	1991	10s	LM	PF	LW	PC	S
Hizballah (Lebanon)	1982–	1,000s	UM	PF	R	RC	—
Hikmatui Zihad (Bangladesh)	2004	10s	L	PF	R	RC	S
Hisba (Nigeria)	2000–	100s	L	PF	R	PC	—
Hizb-I Islami Gulbuddin (Afghanistan, Pakistan)	1977–	100s	L	PF	R	RC	—
Hizbul Mujahideen (Pakistan)	1989–	100s	L	NF	R	TC	—
Holders of the Black Banners (Iraq)	2004–	10s	LM	NF	R	RC	—
HPG (Turkey)	1999–	1,000s	UM	PF	N	TC	—
Iduwini Youths (Nigeria)	1998–	1,000s	L	PF	N	PC	—
Independent Armed Revolutionary Movement (Cuba, Puerto Rico, U.S.)	1956–1971	10s	H	F	N	RC	PO
Indigenous People's Federal Army (Philippines)	2001–	10s	LM	PF	N	RC	—
Inevitables, The (Bolivia)	2003	10s	LM	PF	LW	PC	PT
Informal Anarchist Federation (Italy)	2003–	100s	H	F	LW	SR	—
Ingush Jama'at Shariat (Russia)	2006–	10s	UM	NF	N	TC	—
International Communist Group (Italy)	1984	10s	H	F	LW	SR	S

Table A.1—Continued

Organization	Operating	Peak Size	Econ.[a]	Regime[b]	Type[c]	Goal[d]	Ended[e]
International Justice Group (Egypt, Switzerland)	1995	10s	LM	NF	R	E	S
International Revolutionary Action Group (France)	1974–1975	10s	H	F	LW	SR	S
International Solidarity (Italy)	1990–	10s	H	F	LW	SR	—
Internet Black Tigers (Sri Lanka)	1997	10s	LM	PF	N	TC	S
Iparretarrak (France)	1973–2000	10s	H	F	N	TC	PO
Iraqi Democratic Front (Iraq)	1982–1983	10s	UM	NF	N	RC	PT
Iraqi Legitimate Resistance (Iraq)	2004	10s	LM	NF	N	RC	—
Iraqi Revenge Brigades (Iraq)	2005–	10s	LM	NF	N	RC	—
Irish National Liberation Army (Ireland, Northern Ireland, UK)	1974–1998	100s	H	F	N	TC	PO
Irish Republican Army (Ireland, Northern Ireland, UK)	1919–2005	1,000s	H	F	N	TC	PT
Islami Chhatra Shibir (Bangladesh)	1941–	1,000s	L	PF	R	RC	—
Islami Inqilabi Mahaz (India, Pakistan)	1997–	10s	L	NF	R	TC	—
Islamic Action in Iraq (Iran, Iraq)	1984–1991	10s	LM	NF	R	RC	MF
Islamic Army in Iraq (Iraq)	2003–	1,000s	LM	NF	R	RC	—
Islamic Brotherhood (Guatemala)	1991	10s	LM	PF	LW	PC	PT
Islamic Defense Force (India)	1997–1998	10s	L	F	R	PC	PO

Table A.1—Continued

Organization	Operating	Peak Size	Econ.[a]	Regime[b]	Type[c]	Goal[d]	Ended[e]
Islamic Glory Brigades in the Land of the Nile (Egypt)	2005	10s	LM	NF	R	RC	PO
Islamic Great Eastern Raiders Front (Turkey)	1970–	10s	UM	PF	R	RC	—
Islamic International Peacekeeping Brigade (Azerbaijan, Turkey, Russia, Georgia)	1998–	100s	UM	NF	R	TC	—
Islamic Jihad Brigades (Iraq)	2004–	10s	LM	NF	R	RC	—
Islamic Jihad Group (Uzbekistan)	2004–	10s	L	NF	R	RC	—
Islamic Liberation Organization (Egypt, Lebanon)	1967–1985	10s	LM	PF	R	E	S
Islamic Movement for Change (Saudi Arabia, Syria)	1995–1997	10s	UM	NF	R	E	S
Islamic Movement of Uzbekistan (Afghanistan, Iran, Kazakhstan, Tajikistan, Kyrgyzstan, Pakistan, Uzbekistan)	1998–	100s	L	NF	R	E	—
Islamic Resistance Brigades (Iraq)	2004–	10s	LM	NF	R	RC	—
Islamic Salvation Front (Algeria, Germany, U.S.)	1989–2000	1,000s	LM	NF	R	PC	PT
Islamic Shashantantra Andolon (Bangladesh)	2002–	1,000s	L	PF	R	RC	—
Jagrata Muslim Janata Bangladesh (Bangladesh)	1998–	10,000s	L	PF	R	RC	—
Jaime Bateman Cayon Group (Colombia)	1989–2002	10s	LM	PF	LW	RC	PO
Jaish al-Taifa al-Mansoura (Iraq)	2003–	10s	LM	NF	R	RC	—
Jaish-e-Mohammad (Pakistan)	2000–	100s	L	NF	R	TC	—

Table A.1—Continued

Organization	Operating	Peak Size	Econ.[a]	Regime[b]	Type[c]	Goal[d]	Ended[e]
Jaish-ul-Muslimin (Afghanistan)	2004–	10s	L	PF	R	RC	—
Jamatul Mujahedin Bangladesh (Bangladesh)	2002–	10,000s	L	PF	R	RC	—
Jamiat ul-Mujahedin (Pakistan)	1990–	1,000s	L	NF	R	TC	—
Jammu and Kashmir Islamit Front (Pakistan)	1994–1996	100s	L	PF	N	TC	PO
Janashakti (India)	1992–	100s	L	F	LW	RC	—
January 31 Popular Front (Guatemala)	1981–1982	10s	LM	NF	LW	RC	S
Japanese Red Army (Japan, Lebanon)	1970–2001	10s	H	F	LW	SR	PO
Jemaah Islamiya (Indonesia, Malaysia, Singapore, Philippines)	1993–	100s	LM	PF	R	E	—
Jenin Martyr's Brigade (Israel, West Bank/Gaza)	2003–	10s	LM	PF	N	TC	—
Jewish Defense League (U.S.)	1968–1987	10s	H	F	R	PC	PO
Jihad Committee (India)	1986–	10s	L	F	R	PC	—
Jordanian Islamic Resistance (Jordan)	1997–2000	10s	LM	PF	R	RC	S
July 20th Brigade (Italy)	2002	10s	H	F	LW	SR	S
Jund Allah Organization for the Sunni Mujahideen in Iran (Iran)	2005	10s	LM	NF	R	RC	S
Jund al-Sham (Lebanon, Syria)	1999–	10s	LM	NF	R	E	—
June 16 Organization (Turkey)	1987–1989	10s	LM	PF	LW	SR	PT

Table A.1—Continued

Organization	Operating	Peak Size	Econ.[a]	Regime[b]	Type[c]	Goal[d]	Ended[e]
Justice Army of Defensless People (Mexico)	1996–1998	10s	UM	PF	LW	PC	S
Justice Commandos for the Armenian Genocide (Lebanon, U.S.)	1975–1983	10s	H	F	N	TC	PO
Kabataang Makabayan (Philippines)	1964–	1,000s	LM	PF	LW	RC	—
Kach (Israel, West Bank/Gaza)	1971–	10s	H	F	R	TC	—
Kahane Chai (Israel, West Bank/Gaza)	1990–	10s	H	F	R	TC	—
Kakurokyo (Japan)	1969–	1,000s	H	F	LW	PC	—
Kanglei Yawol Kanna Lup (India)	1994–	100s	L	F	N	SR	—
Kangleipak Communist Party (India)	1980–	10s	L	F	N	TC	—
Karbala Brigades (Iraq)	2004	10s	LM	NF	N	PC	S
Karbi Longri North Cachar Hills Resistance Force (India)	2004–	10s	L	F	N	TC	—
Karenni National Progressive Party (Burma, Thailand)	1955–	1,000s	L	NF	N	TC	—
Kayin National Union (Burma, Thailand)	1959–	1,000s	L	NF	N	TC	—
Kenkoku Giyugun Chosen Seibatsutai (Japan)	2002–2004	10s	H	F	RW	PC	PO
Khaibar Brigades (Lebanon)	1985	10s	LM	PF	N	PC	S
Khmer Rouge (Cambodia)	1951–1998	1,000s	L	NF	LW	RC	MF

Table A.1—Continued

Organization	Operating	Peak Size	Econ.[a]	Regime[b]	Type[c]	Goal[d]	Ended[e]
Khristos Kasimis Revolutionary Group for International Solidarity (Greece)	1985–1986	10s	UM	F	LW	SR	S
Knights of the Tempest (West Bank/Gaza)	2005–	10s	LM	PF	N	TC	—
Komando Jihad (Indonesia)	1975–1981	100s	LM	PF	R	RC	PO
Kosovo Liberation Army (Macedonia, Serbia, Montenegro)	1992–1999	10,000s	LM	PF	N	TC	V
Ku Klux Klan (U.S.)	1866–	1,000s	H	F	RW	SR	—
Kuki Liberation Army (India)	1998–2005	10s	L	F	N	TC	MF
Kuki Revolutionary Army (India)	1999–	10s	L	F	N	TC	—
Kumpulan Mujahidin Malaysia (Indonesia, Malaysia, Philippines)	1995–	100s	UM	PF	R	E	—
Kurdish Islamic Unity Party (Turkey)	1995	10s	LM	PF	N	TC	S
Kurdish Patriotic Union (Turkey)	1994	10s	LM	PF	N	TC	S
Kurdistan Freedom Hawks (Iraq, Turkey)	2004–	100s	UM	PF	N	TC	—
Kurdistan Workers' Party (Turkey)	1974–	1,000s	UM	PF	N	TC	—
Lashkar-e-Jabbar (India)	2001–	10s	L	F	R	PC	—
Lashkar-e-Jhangvi (India, Pakistan)	1996–	100s	L	NF	R	TC	—
Lashkar-e-Taiba (India, Pakistan)	1989–	1,000s	L	NF	R	TC	—

Table A.1—Continued

Organization	Operating	Peak Size	Econ.[a]	Regime[b]	Type[c]	Goal[d]	Ended[e]
Lashkar-I-Omar (Pakistan)	2001–	10s	L	NF	R	RC	—
Laskar Jihad (Indonesia)	2000–	10,000s	LM	F	R	RC	—
Latin American Patriotic Army (Colombia, Venezuela)	2001	10s	UM	PF	LW	RC	S
Lautaro Youth Movement (Chile)	1983–1994	100s	UM	F	LW	RC	PO
Lebanese Arab Youth (Lebanon)	1977	10s	LM	PF	N	PC	S
Lebanese Armed Revolutionary Faction (Lebanon)	1979–1986	100s	LM	PF	LW	RC	PO
Lebanese Liberation Front (Lebanon)	1987–1989	10s	LM	NF	N	PC	PO
Lebanese National Resistance Front (Lebanon)	1982–1990	100s	LM	NF	LW	RC	S
Lebanese Socialist Revolutionary Organization (Lebanon)	1973–1974	10s	LM	F	LW	PC	PO
Liberation Front of Quebec (Canada)	1963–1972	100s	H	F	N	TC	PO
Liberation Tigers of Tamil Eelam (Sri Lanka)	1976–	10,000s	LM	PF	N	TC	—
Libyan Islamic Fighitng Group (Libya)	1995–	100s	UM	NF	R	RC	—
Lord's Resistance Army (Democratic Republic of the Congo, Sudan, Uganda)	1992–	1,000s	L	PF	R	RC	—
Los Macheteros (Puerto Rico, U.S.)	1976–1999	10s	UM	F	N	RC	PO
Loyalist Volunteer Force (Northern Ireland, UK)	1997–2003	100s	H	F	N	SQ	PT

Table A.1—Continued

Organization	Operating	Peak Size	Econ.[a]	Regime[b]	Type[c]	Goal[d]	Ended[e]
Macedonian Revolutionary Organization (Macedonia, Greece)	2001	1,000s	LM	PF	N	TC	PT
Mahdi Army (Iraq)	2003–	1,000s	LM	NF	N	PC	—
Mahir Cayan Suicide Group (Turkey)	1975	10s	LM	F	LW	RC	S
Manuel Rodriguez Patriotic Front (Chile)	1989–1994	100s	UM	F	LW	PC	PO
Manuel Rodriguez Patriotic Front (Chile)	1983–1989	1,000s	LM	PF	LW	RC	V
Maoist Communist Center (India)	1969–2004	1,000s	L	F	LW	TC	S
Mariano Moreno National Liberation Comando (Argentina)	2005	10s	UM	F	LW	PC	PT
Martyr Abu-Ali Mustafa Brigades (Israel, West Bank/Gaza)	2001–	10s	LM	PF	N	TC	—
Masada, Action and Defense Movement (France)	1972–1988	10s	H	F	RW	SR	PO
Maximiliano Gomez Revolutionary Brigade (Dominican Republic)	1987–1987	10s	LM	F	LW	PC	PT
May 15 Organization for the Liberation of Palestine (Iraq)	1979–1985	10s	UM	NF	N	RC	S
May 19 Communist Order (U.S.)	1983–1986	10s	H	F	LW	PC	PO
Meinhof-Puig-Antich Group (France)	1975	10s	H	F	LW	SR	PT
Mohammed's Army (Yemen)	2000	10s	L	NF	R	RC	PO
Montoneros (Argentina)	1970–1981	10s	UM	PF	LW	RC	MF

Table A.1—Continued

Organization	Operating	Peak Size	Econ.[a]	Regime[b]	Type[c]	Goal[d]	Ended[e]
Morazanist Front for the Liberation of Honduras (Honduras, Nicaragua)	1980–1992	100s	L	F	LW	RC	PT
Morazanist Patriotic Front (Honduras)	1988–1995	10s	L	PF	LW	PC	MF
Moro Islamic Liberation Front (Philippines)	1977–2001	100s	LM	F	R	TC	PT
Moro National Liberation Front (Philippines)	1972–	10,000s	LM	PF	R	TC	—
Moroccan Islamic Combatant Group (Morocco, Afghanistan, Europe)	1990–	100s	LM	PF	R	RC	—
Movement for Democracy and Development (Cameroon, Chad, Libya, Nigeria)	1991–2003	1,000s	L	NF	LW	RC	S
Movement for Democracy and Justice in Chad (Chad, Libya)	1998–2003	100s	L	NF	LW	RC	PT
Movement for the Emancipation of the Niger Delta (Niger)	2006–	1,000s	L	PF	N	PC	—
Movement of the Islamic Action of Iraq (Iran, Iraq)	1982	10s	UM	NF	R	RC	S
Movement of the Revolutionary Left (Chile)	1965–2004	100s	UM	F	LW	RC	PT
Movimiento Armado Nacionalista Organizacion (Mexico, Peru)	1974	10s	LM	NF	RW	SQ	PT
Movsar Baryayev Gang (Russia)	1998–2002	10s	LM	PF	N	TC	PO
Mozambique National Reisistance Movement (Mozambique, Rhodesia, South Africa)	1976–1992	1,000s	L	PF	RW	SQ	PT
Mujahedin-e-Khalq (France, Iraq)	1971–	100s	LM	NF	N	RC	—

Table A.1—Continued

Organization	Operating	Peak Size	Econ.[a]	Regime[b]	Type[c]	Goal[d]	Ended[e]
Mujahideen al-Mansooran (India)	2002	10s	L	F	N	TC	S
Mujahideen Division Khandaq (Indonesia, Malaysia)	2000	10s	L	PF	R	TC	S
Mujahideen KOMPAK (Indonesia)	2001–	100s	LM	F	R	RC	—
Mujahideen Message (Afghanistan)	2003–	10s	L	PF	R	RC	—
Mujahideen Shura Council (Iraq)	2005–	1,000s	LM	NF	R	RC	—
Muslim United Army (Pakistan)	2002–2003	10s	L	NF	R	RC	S
Muslims Against Global Oppression (South Africa)	1998	100s	LM	F	R	PC	S
Muttahida Qami Movement (Pakistan)	1978–2001	1,000s	L	NF	N	PC	PT
Nation of Yahweh (U.S.)	1979–1995	100s	H	F	R	SR	PO
National Anti-Corruption Front (Bolivia)	2005–	10s	LM	PF	LW	PC	—
National Army for the Liberation of Uganda (Democratic Republic of the Congo, Uganda)	1988–	1,000s	L	PF	N	RC	—
National Democratic Front of Bodoland (Bhutan, Burma, India)	1988–	1,000s	L	F	N	TC	—
National Front for the Liberation of Angola (Angola, Democratic Republic of the Congo)	1962–1990	1,000s	LM	NF	N	RC	MF
National Front for the Liberation of Kurdistan (France, Greece, Russia)	1985–	1,000s	H	F	N	TC	—

Table A.1—Continued

Organization	Operating	Peak Size	Econ.[a]	Regime[b]	Type[c]	Goal[d]	Ended[e]
National Liberation Army (the first) (Bolivia)	1966–1970	100s	LM	PF	LW	RC	MF
National Liberation Army (the second) (Bolivia)	1987–2003	100s	LM	F	LW	RC	PO
National Liberation Army (Colombia)	1964–	1,000s	LM	PF	LW	RC	—
National Liberation Front of Tripura (Bangladesh, India)	1989–	1,000s	L	F	N	TC	—
National Liberation Union (Suriname)	1989	10s	UM	PF	LW	PC	S
National Patriotic Front of Liberia (Cote d'Ivoire, Liberia, Libya)	1984–1995	1,000s	L	NF	LW	RC	V
National Revolutionary Command (Lebanon)	1986	10s	LM	PF	N	PC	S
National Socialist Council of Nagaland-Isak-Muivah (India)	1988–	1,000s	L	F	N	TC	—
National Socialist Council of Nagaland-Khaplang (Burma, India)	1988–	1,000s	L	F	N	TC	—
National Union for the Total Independence of Angola (Angola)	1966–2002	1,000s	L	NF	LW	RC	PT
National Warriors (South Africa)	2002	10s	LM	F	RW	TC	PT
National Youth Reistance Organization (Italy)	1973	10s	H	F	LW	RC	S
Nationalist Kurdish Revenge Teams (Turkey)	1999	10s	LM	PF	N	TC	S
Nestor Paz Zamora Commission (Bolivia)	1990–1991	10s	LM	F	LW	RC	S

Table A.1—Continued

Organization	Operating	Peak Size	Econ.[a]	Regime[b]	Type[c]	Goal[d]	Ended[e]
New Armenian Resistance (Belgium, France, Italy, UK, Armenia)	1977–1983	10s	H	F	N	TC	S
New People's Army (Philippines)	1969–	10,000s	LM	PF	LW	RC	—
New Red Brigades/Communist Combatant Party (Italy)	1984–	10s	H	F	LW	RC	—
New Revolutionary Alternative (Russia)	1996–2001	10s	LM	PF	LW	RC	PO
Night Avengers (Honduras)	1997–1998	10s	L	F	RW	PC	PT
Ninth of June Organization (France)	1981–1982	10s	H	F	N	PC	S
November's Children (Greece)	1996–2001	10s	H	F	LW	SR	S
Nuclei Armati Comunista (Italy)	1982	10s	H	F	LW	RC	S
Nuclei Communist Combatants (Italy)	1994	10s	H	F	LW	RC	PT
Nusantara Islamic Jihad Forces (Indonesia)	1999	100s	L	PF	R	E	S
Odua Peoples' Congress (Nigeria)	1995–	100s	L	PF	N	TC	—
Ogaden National Liberation Front (Ethiopia, Somalia)	1984–	100s	L	PF	N	TC	—
Oklahoma City Bombing Conspirators (U.S.)	1995	10s	H	F	RW	RC	PO
Omar Torrijos Commando for Latin American Dignity (Panama, Venezuela)	1989–1990	10s	UM	F	N	PC	V
Omega-7 (U.S.)	1974–1983	10s	H	F	N	RC	PO

Table A.1—Continued

Organization	Operating	Peak Size	Econ.[a]	Regime[b]	Type[c]	Goal[d]	Ended[e]
OPR-33 (Argentina, Uruguay)	1971–1976	10s	UM	NF	LW	SR	MF
Orange Volunteers (Northern Ireland)	1998–2001	10s	H	F	N	SQ	PO
Order, The (U.S.)	1982–1984	10s	H	F	RW	SR	PO
Order II, The (U.S.)	1986	10s	H	F	RW	SR	PO
Orly Organization (France)	1981–1983	10s	H	F	N	PC	S
Oromo Liberation Front (Ethiopia, Somalia	1973–	1,000s	L	PF	N	TC	—
Padanian Armed Separatist Phalanx (Italy)	1998	10s	H	F	N	TC	PT
Palestine Liberation Front (Iraq, Lebanon, Libya, Tunisia)	1959–1996	100s	LM	NF	N	TC	PT
Palestine Liberation Organization (West Bank/Gaza)	1964–	10,000s	LM	PF	N	TC	—
Palestine Resistance (France)	1980	10s	H	F	N	PC	S
Palestinian Islamic Jihad (Israel, Syria, West Bank/Gaza)	1978–	100s	LM	PF	R	TC	—
Palestinian Popular Struggle Front (Lebanon, Syria)	1967–1991	10s	LM	NF	N	TC	PT
Palestinian Revolution Forces General Command (Israel, West Bank/Gaza)	1985–1987	100s	LM	NF	N	TC	S
Pan-Turkish Organization (Bulgaria, Netherlands)	1985	10s	UM	NF	N	RC	S
Parbatya Chattagram Jana Sanghati Samity (Bangladesh)	1972–	100s	L	PF	N	TC	—

Table A.1—Continued

Organization	Operating	Peak Size	Econ.[a]	Regime[b]	Type[c]	Goal[d]	Ended[e]
Partisans of Holy War (Lebanon)	1987	10s	LM	NF	R	PC	S
Pattani United LIberation Organization (Malaysia, Thailand)	1968–	10s	LM	NF	R	TC	—
Peace Conquerors (Austalia, Belgium, Germany)	1985	10s	H	F	LW	PC	PT
Pedro Leon Arboleda Movement (Colombia)	1979–1987	10s	LM	F	LW	RC	S
People Against Gangsterism and Drugs (South Africa)	1995–	100s	UM	F	RW	PC	—
People's Command (Bolivia)	1986	10s	LM	F	LW	PC	PT
People's Liberation Army (Bangladesh, Burma, India)	1978–	1,000s	L	F	N	TC	—
People's Liberation Army of Kurdistan (Turkey)	1985–	1,000s	UM	PF	N	TC	—
People's Liberation Forces (Colombia)	1998–2000	10s	LM	PF	LW	RC	PO
People's Liberation Forces (El Salvador)	1970–1991	100s	LM	PF	LW	RC	PT
People's Revolutionary Armed Forces (Mexico)	1972–1977	10s	LM	PF	LW	RC	S
People's Revolutionary Army (Argentina)	1960–1977	100s	UM	NF	LW	RC	MF
Peoples Revolutionary Front (Philippines)	1971	10s	LM	PF	LW	PC	S
People's Revolutionary Militias (Ecuador)	2003–	10s	LM	PF	LW	PC	—
People's Revolutionary Organization (Argentina)	1992–1997	10s	UM	F	LW	RC	PO

Table A.1—Continued

Organization	Operating	Peak Size	Econ.[a]	Regime[b]	Type[c]	Goal[d]	Ended[e]
People's Revolutionary Party (Democratic Republic of the Congo, Tanzania, Zaire [Belgian Congo])	1967–1997	1,000s	L	NF	LW	RC	V
People's United Liberation Front (India)	1995–	100s	L	F	R	TC	—
People's War Group (India)	1980–2004	100s	L	F	LW	TC	S
Peronist Armed Forces (Argentina)	1967–1974	10s	UM	PF	LW	RC	MF
Peykar (Iran, Switzerland)	1975–1982	10s	UM	NF	LW	RC	S
Phineas Priests (U.S.)	1990–	10s	H	F	R	SR	—
Polisario Front (Algeria, Mauritania, Morocco, Western Sahara)	1973–2005	10,000s	LM	PF	N	TC	PT
Polish Revolutionary Home Army (Poland, Switzerland)	1982	10s	UM	PF	LW	PC	PO
Popular Forces of April 25 (Portugal)	1981–1986	10s	UM	F	LW	RC	PO
Popular Front for the Liberation of Palestine, General Command (Lebanon, Syria)	1968–	100s	LM	NF	N	TC	—
Popular Front for the Liberation of Palestine (Israel, West Bank/Gaza)	1967–	100s	LM	PF	N	TC	—
Popular Liberation Army (Colombia)	1967–	1,000s	LM	PF	LW	RC	—
Popular Movement for the Liberation of Angola (Angola)	1956–1975	10,000s	L	NF	N	RC	V
Popular Resistance Committees (Israel, West Bank/Gaza)	2000–	10s	LM	PF	N	TCs	—

Table A.1—Continued

Organization	Operating	Peak Size	Econ.[a]	Regime[b]	Type[c]	Goal[d]	Ended[e]
Popular Revolutionary Action (Greece)	2003–2005	10s	H	F	LW	SR	PO
Popular Revolutionary Army (Mexico)	1996–	100s	UM	F	LW	RC	—
Popular Revolutionary Resistance Group (Greece)	1971–1972	10s	UM	NF	N	RC	V
Popular Revolutionary Vanguard (Brazil)	1968–1973	100s	LM	PF	LW	RC	MF
Popular Self-Defense Forces (Democratic Republic of the Congo)	1993–	1,000s	L	NF	N	SQ	—
Proletarian Combatant Groups (Italy)	2004–	10s	H	F	LW	SR	—
Proletarian Nuclei for Communism (Italy)	2003–	10s	H	F	LW	SR	—
Protectors of Islam Brigade (Iraq)	2005	10s	LM	NF	R	RC	S
Pueblo Reagrupado (Chile)	2002	10s	UM	F	LW	PC	PT
Puerto Rican Resistance Movement (Puerto Rico, U.S.)	1981	10s	H	F	N	RC	S
Purbo Banglar Communist Party (Bangladesh)	2002–	100s	L	PF	LW	RC	—
Rajah Solaiman Movement (Philippines)	2002–	100s	LM	PF	R	RC	—
Raul Sendic International Brigade (France, Uruguay)	1974	10s	H	F	LW	SR	PO
Real Irish Republican Army (Ireland, Northern Ireland, UK)	1998–	100s	H	F	N	TC	—
Rebel Armed Forces (Guatemala)	1962–1996	1,000s	LM	PF	LW	RC	PT

Table A.1—Continued

Organization	Operating	Peak Size	Econ.[a]	Regime[b]	Type[c]	Goal[d]	Ended[e]
Rebolusyonaryong Hukbong Bayan (Philippines)	1998–	100s	LM	PF	LW	RC	—
Recontra 380 (Honduras, Nicaragua)	1993–1997	100s	L	PF	RW	PC	PT
Red Army Faction (Germany)	1978–1992	10s	H	F	LW	SR	PO
Red Brigades (Italy)	1969–1984	100s	H	F	LW	RC	PO
Red Daughters of Rage (Austria)	1995	10s	H	F	LW	SR	S
Red Flag (Venezuela)	1969–1998	100s	UM	F	LW	RC	PO
Red Guerrillas (Russia)	2002	10s	LM	PF	LW	RC	PT
Red Hand Defenders (Northern Ireland, UK)	1998–	100s	H	F	N	SQ	—
Republic of New Africa (U.S.)	1968–1971	100s	H	F	N	TC	PO
Republic of Texas (U.S.)	1995–1998	100s	H	F	RW	TC	PO
Resistenza Corsa (France)	2002–2003	10s	H	F	N	TC	S
Revenge of the Hebrew Babies (Israel, West Bank/Gaza)	2002–2003	10s	LM	PF	N	SQ	PT
Revolutionary Action Party (U.S.)	1970	10s	H	F	LW	PC	PT
Revolutionary Armed Forces of Colombia (Colombia)	1964–	10,000s	LM	PF	LW	RC	—
Revolutionary Armed Forces of the People (Mexico)	2000–	10s	UM	F	LW	RC	—
Revolutionary Army (Japan)	2000–	10s	H	F	LW	PC	—

Table A.1—Continued

Organization	Operating	Peak Size	Econ.[a]	Regime[b]	Type[c]	Goal[d]	Ended[e]
Revolutionary Army of the People (Dominican Republic)	1989	10s	LM	F	LW	PC	PT
Revolutionary Autonomous Group (Portugal)	1985	10s	UM	F	LW	RC	S
Revolutionary Cells Animal Liberation Brigade (U.S.)	2003	10s	H	F	LW	PC	PO
Revolutionary Commandos of Solidarity (Costa Rica)	1977	10s	LM	F	LW	PC	PT
Revolutionary Eelam Organization (Sri Lanka)	1975–1990	100s	L	PF	N	TC	S
Revolutionary Force Seven (U.S.)	1970	10s	H	F	LW	PC	PT
Revolutionary Front for Communism (Italy)	1996–	10s	H	F	LW	RC	—
Revolutionary Front for Proletarian Action (Belgium)	1985	10s	H	F	LW	SR	PO
Revolutionary Leninist Brigades (Italy)	2000	10s	H	F	LW	RC	S
Revolutionary Movement of October 8 (Brazil)	1968–1972	100s	LM	PF	LW	RC	PO
Revolutionary Nuclei (Greece)	1974–2003	10s	H	F	LW	SR	S
Revolutionary Offensive Cells (Italy)	2003–	10s	H	F	LW	SR	—
Revolutionary Organization 17 November (Greece)	1975–2002	10s	H	F	LW	SR	PO
Revolutionary Organization of Socialist Muslims (Iraq, Libya, Syria)	1974–1985	10s	UM	NF	N	TC	S
Revolutionary People's Front (Bangladesh)	1979–	1,000s	L	PF	N	TC	—

Table A.1—Continued

Organization	Operating	Peak Size	Econ.[a]	Regime[b]	Type[c]	Goal[d]	Ended[e]
Revolutionary People's Struggle (Greece)	1975–1995	10s	UM	F	LW	SR	S
Revolutionary Perspective (Spain)	2000	10s	H	F	LW	RC	S
Revolutionary Proletarian Initiative Nuclei (Italy)	2000–	10s	H	F	LW	SR	—
Revolutionary Proletarian Nucleus (Italy)	2000–	10s	H	F	LW	SR	—
Revolutionary Socialists (Sweden)	1999–2000	10s	H	F	LW	PC	PT
Revolutionary Struggle (Greece)	2003–	10c	H	F	LW	SR	—
Revolutionary United Front (Liberia, Sierra Leone)	1987–2002	10,000s	L	PF	LW	RC	MF
Revolutionary United Front Movement (Honduras)	1989	10s	LM	F	LW	PC	V
Revolutionary Worker Clandestine Union of the People Party (Mexico)	1970–	100s	UM	F	LW	RC	—
Riyad-us-Saliheyn Martyrs' Brigade (Russia)	2002–	100s	UM	NF	N	TC	—
Russian National Bolshevist Party (Estonia, Kyrgyzstan, Moldova, Russia, Latvia)	1993–	10,000s	UM	NF	LW	SR	—
Russian National Unity (Estonia, Latvia, Lithuania, Ukraine, Russia)	1990–	10,000s	UM	NF	RW	SR	—
Saif-ul-Muslimeen (Afghanistan)	2003–	10s	L	PF	R	RC	—
Salafia Jihadia (Morocco)	1996–	100s	LM	PF	R	E	—

Table A.1—Continued

Organization	Operating	Peak Size	Econ.[a]	Regime[b]	Type[c]	Goal[d]	Ended[e]
Salafist Group for Call and Combat (Algeria, Mauritania, Mali, Niger)	1996–	100s	LM	NF	R	RC	—
Salah al-Din Battalions (Israel, West Bank/Gaza)	2000–	10s	LM	PF	N	TC	—
Sandinistas (Nicaragua)	1960–1979	10,000s	LM	PF	LW	RC	V
Saraya al-Shuhuada al-Jihadiyah fi al-Iraq (Iraq)	2004–	10s	LM	NF	R	RC	—
Sardinian Autonomy Movement (Italy)	2002	10s	H	F	N	TC	PT
Save Kashmir Movement (India)	2002–	10s	L	F	N	TC	—
Secret Army Organization (U.S.)	1969–1972	10s	H	F	RW	SQ	PO
Secret Organization Zero (U.S.)	1974–1975	10s	H	F	RW	RC	PT
Sekihotai (Japan)	1987–1990	10s	H	F	RW	PC	PT
Self-Defense Groups of Cordoba and Uraba (Colombia)	1994–2006	1,00s	LM	PF	RW	SQ	PT
September-France (France)	1981	10s	H	F	N	TC	S
Shahin (Iran)	1992	10s	LM	NF	N	TC	PT
Shining Path (Peru)	1980–	1,000s	LM	F	LW	RC	—
Shurafa al-Urdun (Jordan)	2001–2002	10s	LM	PF	R	E	PO
Simon Bolivar Guerilla Coordinating Board (Colombia)	1987–1997	10,000s	LM	PF	LW	RC	S

Table A.1—Continued

Organization	Operating	Peak Size	Econ.[a]	Regime[b]	Type[c]	Goal[d]	Ended[e]
Sipah-e-Sahaba Pakistan (Pakistan)	1985–	100s	L	NF	R	RC	—
Socialist-Nationalist Front (Switzerland)	1988	10s		F	LW	SR	PT
Sons of the South (Lebanon)	1984	10s	LM	PF	N	PC	S
South Londonderry Volunteers (Northern Ireland, UK)	1998–2001	10s	H	F	N	SQ	S
South Maluku Republic (Indonesia, Netherlands)	1998–	100s	LM	PF	R	TC	—
South Moluccan Suicide Commando (Netherlands)	1978	10s	H	F	N	TC	PO
Southern Sudan Independence Movement (Sudan)	1991–1996	1,000s	L	NF	N	TC	PT
South-West Africa People's Organization (Angola, Namibia)	1960–1989	1,000s	LM	PF	N	RC	V
Sovereign Panama Front (Panama)	1992–1999	100s	UM	F	N	PC	V
Spanish Basque Battalion (France, Spain)	1975–1982	10s	H	F	RW	SQ	S
Spanish National Action (France, Spain)	1979	10s	H	F	RW	SQ	S
Special Purpose Islamic Regiment (Georgia, Russia)	1996–	1,000s	UM	NF	N	TC	—
Sri Nakharo (Malaysia, Thailand)	2001	10s	LM	F	R	TC	PO
Strugglers for the Unity and Freedom of Greater Syria (Lebanon)	2005	10s	UM	PF	N	RC	S
Students Islamic Movement of India (India)	1977–	100s	L	F	R	E	—

Table A.1—Continued

Organization	Operating	Peak Size	Econ.[a]	Regime[b]	Type[c]	Goal[d]	Ended[e]
Sudan People's Liberation Army (Sudan)	1983–2005	10,000s	L	NF	N	PC	PT
Supporters of Horst Ludwig Meyer (Denmark)	1999	10s	H	F	LW	PC	PT
Sword of Islam (Russia)	1998–2001	10s	LM	PF	N	TC	S
Swords of Righteousness Brigades (Iraq)	2005–	10s	LM	NF	R	RC	—
Syrian Social Nationalist Party (Lebanon, Syria)	1932–2005	10,000s	LM	NF	N	E	PT
Takfir wa Hijra (Algeria, Egypt, France, Germany, Italy, Lebanon, Morocco, Netherlands, Spain, UK)	1971–	100s	LM	NF	R	E	—
Taliban (Afghanistan)	1994–	1,000s	L	PF	R	RC	—
Tanzim (Israel, West Bank/Gaza)	1993–	100s	LM	PF	N	TC	—
Taong Bayan at Kawal (Philippines)	2006–	10s	LM	PF	LW	RC	—
Tawhid and Jihad (Iraq)	1999–	100s	LM	NF	R	RC	—
Tawhid Islamic Brigades (Egypt)	2004	10s	LM	NF	R	PC	S
Tera Lliure (Spain)	1972–1991	100s	H	F	N	TC	PO
Territorial Anti-Imperialist Nuclei (Italy)	1995–	10s	H	F	LW	RC	—
Third of October Group (France, Switzerland)	1980–1981	10s	H	F	N	RC	S
Tigers (Swaziland)	1998	10s	LM	NF	N	RC	PO

Table A.1—Continued

Organization	Operating	Peak Size	Econ.[a]	Regime[b]	Type[c]	Goal[d]	Ended[e]
Tigray Peoples Liberation Front (Ethiopia)	1975–1991	1,000s	L	PF	LW	RC	PT
TKEP/L (Turkey)	1990–2001	100s	LM	PF	LW	TC	PO
TKP/ML-TIKKO (Turkey)	1972–	100s	UM	PF	LW	RC	—
Tontons Macoutes (Haiti)	1959–2000	100s	L	NF	RW	RC	PO
Totally Anti-War Group (France)	2001	10s	H	F	LW	PC	PT
Tunisian Combatant Group (Afghanistan, Tunisia, Western Europe)	2000–	10s	LM	NF	R	RC	—
Tupac Amaru Revolutionary Movement (Peru)	1982–	10s	LM	F	LW	RC	—
Tupac Katari Guerrilla Army (Bolivia)	1991–1993	100s	LM	F	LW	RC	PO
Tupamaro Revolutionary Movement, January 23 (Venezuela)	1998–2003	10s	UM	PF	LW	SQ	PT
Tupamaros (Uruguay)	1963–1985	1,000s	UM	F	LW	RC	MF
Turkish Hizballah (Turkey)	1981–	100s	UM	PF	R	RC	—
Turkish Islamic Jihad (Turkey)	1991–1996	100s	LM	PF	R	RC	S
Turkish People's Liberation Army (Turkey)	1971–1980	10s	LM	PF	LW	RC	PO
Turkish People's Liberation Front (Turkey)	1971–1999	100s	LM	PF	LW	RC	S
Uganda Democratic Christian Army (Sudan, Uganda)	1990–1994	100s	L	PF	R	RC	S

Table A.1—Continued

Organization	Operating	Peak Size	Econ.[a]	Regime[b]	Type[c]	Goal[d]	Ended[e]
Ulster Defence Association/Ulster Freedom Fighters (Northern Ireland, UK)	1971–	10,000s	H	F	N	SQ	—
Ulster Volunteer Force (Northern Ireland, UK)	1966–	1,000s	H	F	N	SQ	—
Ummah Liberation Army (Sudan)	1990–2000	100s	L	NF	LW	RC	PT
Underground Government of the Free Democratic People of Laos (Laos)	2000–	100s	L	NF	N	RC	—
United Anti-Reelection Command (Dominican Republic)	1970	10s	LM	F	LW	PC	S
United Arab Revolution (Kuwait)	1986	10s	H	PF	N	E	PO
United Freedom Front (U.S.)	1974–1984	10s	H	F	LW	PC	PO
United Kuki Liberation Front (India)	1999–	10s	L	F	N	TC	—
United Liberation Front of Assam (India)	1979–	1,000s	L	F	N	TC	—
United Nasserite Organization (Cyprus, Lebanon)	1986–1987	10s	UM	F	N	PC	S
United National Liberation Front (Bangladesh, Burma, India)	1990–	1,000s	L	F	LW	TC	—
United Organization of Halabjah Martyrs (Iraq)	1989	10s	UM	NF	N	PC	S
United People's Democratic Front (Bangladesh)	1998–	100s	L	PF	N	TC	—
United People's Democratic Solidarity (India)	1999–	100s	L	F	N	TC	—
United Popular Action Movement (Chile)	1986–1992	100s	LM	F	LW	RC	PT

Table A.1—Continued

Organization	Operating	Peak Size	Econ.[a]	Regime[b]	Type[c]	Goal[d]	Ended[e]
United Revolutionary Front (Venezuela)	1997–1999	10s	UM	PF	LW	PC	V
United Self-Defense Forces of Colombia (Colombia)	1997–2006	10,000s	LM	PF	RW	SQ	PT
United Self-Defense Forces of Venezuela (Venezuela)	2002–	100s	UM	PF	RW	RC	—
United Tajik Opposition (Tajikistan)	1994–2003	100s	L	NF	LW	RC	PT
Up the IRS, Inc. (U.S.)	1986–1991	10s	H	F	LW	PC	PO
Usd Allah (Iraq)	2004	10s	LM	NF	R	RC	S
Uygur Holy War Organization (China)	2001	10s	LM	NF	N	TC	S
VAR-Palmares (Brazil)	1968–1972	100s	LM	PF	LW	RC	MF
Venceremos (Venezuela)	1988–1991	10s	UM	F	LW	PC	PT
Vigorous Burmese Student Warriors (Burma, Thailand)	1999–	100s	L	NF	LW	RC	—
Vitalunismo (France, Italy)	1999–2002	10s	H	F	N	E	PO
Waffen SS (Latvia)	1998	10s	LM	F	RW	SR	PT
Weather Underground Organization/Weathermen (U.S.)	1969–1977	10s	H	F	LW	PC	PO
West Nile Bank Front (Democratic Republic of the Congo, Uganda)	1995–2004	10,000s	L	PF	LW	RC	MF
White Legion (Ecuador)	2001–2003	100s	LM	PF	RW	SR	S

Table A.1—Continued

Organization	Operating	Peak Size	Econ.[a]	Regime[b]	Type[c]	Goal[d]	Ended[e]
Workers' Revolutionary Party (Bolivia, Peru)	1988	10s	LM	F	LW	RC	S
World Islamic Jihad Group (Yemen)	1998	10s	L	NF	R	RC	S
World Punishment Organization (Switzerland)	1982	10s	H	F	N	TC	PT
World United Formosans for Indepdence (Japan, Taiwan, U.S.)	1970	10s	H	F	N	TC	PT
Yanikian Commandos (U.S.)	1973	10s	H	F	N	PC	V
Yemen Islamic Jihad (Afghanistan, Libya, Yemen, UK, U.S.)	1990–	100s	L	PF	R	PC	—
Young Liberators of Pattani (Thailand)	2002	10s	LM	F	N	TC	S
Youth Action Group (France)	1974–1977	10s	H	F	LW	SR	PT
Zarate Willka Armed Forces of Liberation (Bolivia)	1989	10s	LM	F	LW	PC	PT
Zimbabwe African Nationalist Union (Mozambique, Rhodesia, Zimbabwe)	1965–1980	1,000s	LM	PF	N	RC	V
Zionist Action Group (Portugal)	1982	10s	UM	F	R	PC	S
Zomi Revolutionary Army (India)	1997–	100s	L	F	N	TC	—

Table A.1—Continued

Organization	Operating	Peak Size	Econ.[a]	Regime[b]	Type[c]	Goal[d]	Ended[e]

[a] Designations correspond to the World Bank analytical classification for each terrorist group's base country of operation in the year in which the use of terror ended. Current World Bank classifications have been employed for terrorist groups deemed active at the end of 2006. In 2006, the World Bank used the following thresholds for classifying national economies (in 2005 gross national income [GNI] per capita, calculated using the World Bank Atlas Method): L = low income ($875 or less); LM = lower middle income ($876–$3,465); UM = upper middle income ($3,466–$10,725); and H = high income ($10,726 or more).

[b] Designations correspond to the Freedom House classification of each terrorist group's base country of operation in the year in which the use of terror ended. Current Freedom House classifications have been employed for terrorist groups deemed active at the end of 2006. The status designations of F (free), PF (partly free), and NF (not free) are determined by the combination of the political-rights and civil-liberties ratings for each country or territory. These categories contain numerical ratings between 1 and 7 for each country or territory, with 1 representing the freest and 7 the least free. The aggregate score of 2 to 14 indicates the general state of freedom in each country or territory.

[c] R = religious. LW = left-wing. N = nationalist. RW = right-wing.

[d] RC = regime change. TC = territorial change. PC = policy change. E = empire. SR = social revolution. SQ = status quo.

[e] PO = policing. S = splintering. PT = politics. V = victory. MF = military force.

APPENDIX B

Al Qa'ida Attacks, 1994–2007

We have excluded from this list terrorist incidents in Afghanistan and Iraq, as well as incidents perpetrated by individuals or groups inspired but not directly affiliated with al Qa'ida. In some cases, the affiliation between al Qa'ida and the perpetrators of the incidents on this list cannot be said to be definitive but rather is probable.

Table B.1
Al Qa'ida Attacks, 1994–2007

Date	Location	Description
November 13, 1995	Riyadh, Saudi Arabia	A van packed with plastic explosives exploded outside the Office of the Program Manager of the Saudi Arabian National Guard run by the U.S. military, killing seven and wounding 60.
November 17, 1997	Luxor, Egypt	At least six gunmen dressed as security personnel opened fire on a group of tourists at the Temple of Hatshepsut, killing 68 and wounding 24.
August 7, 1998	Nairobi, Kenya	A suicide car bomb exploded outside the U.S. embassy, killing at least 213 and injuring more than 5,000.
	Dar es Salaam, Tanzania	A suicide car bomb exploded outside the U.S. embassy, killing at least 11 and injuring 77.
December 24, 1999	Delhi, India	Five armed Pakistani men hijacked Indian Airlines flight 814 as it took off from Kathmandu en route to Delhi, killing one.
October 12, 2000	Aden, Yemen	A small boat that was helping the U.S. Navy destroyer USS *Cole* to moor exploded as the ship was in the Yemen port of Aden for refueling, killing 17 U.S. Navy personnel and wounding 39 others.

Table B.1—Continued

Date	Location	Description
September 11, 2001	Washington, D.C., USA	American Airlines flight 77 bound for Los Angeles, California, from Dulles International Airport was hijacked and crashed into the helicopter-landing pad on the west side of the Pentagon, killing 189, wounding 76, and causing extensive damage to the structure of the Pentagon.
	Shanksville, Pa., USA	United Airlines flight 93 bound for San Francisco, California, from Newark, New Jersey, was hijacked and crashed near Shanksville, Pennsylvania, shortly after it turned in the direction of Washington, D.C., killing all 44 passengers on board.
	New York, N.Y., USA	American Airlines flight 11 bound for Los Angeles, California, from Boston, Massachusetts, was hijacked and crashed into the north tower of the World Trade Center. The exact number of casualties that this attack caused is unclear; the combined casualties of both World Trade Center attacks were 2,823 deaths plus hundreds of additional injuries.
September 11, 2001	New York, N.Y., USA	United Airlines flight 175 bound for Los Angeles, California, from Boston, Massachusetts, crashed into the south tower of the World Trade Center. The exact number of casualties that this attack caused is unclear; the combined casualties of both World Trade Center attacks were 2,823 deaths plus hundreds of additional injuries.
April 11, 2002	Djerba Island, Tunisia	A natural-gas truck exploded next to an historic synagogue on the island of Djerba off the coast of Tunisia, killing 15 and wounding 20.
October 28, 2002	Amman, Jordan	Zarqawi's al Qa'ida network killed Laurence Foley, an officer for USAID in Amman, Jordan.
November 28, 2002	Mombasa, Kenya	A shoulder-launched rocket or missile was fired at an Israeli Arkya 757 civilian airliner as it departed Mombasa for Tel Aviv; the rocket missed its target.
		A suicide bomber blew himself up in the lobby of the Israeli-owned Paradise Hotel, while two more suicide attackers detonated a car bomb outside the hotel, killing 13 and wounding 80.

Table B.1—Continued

Date	Location	Description
May 12, 2003	Riyadh, Saudi Arabia	A suicide bomber detonated a vehicle within the Al Hamra housing complex, a residence for foreigners working in Saudi Arabia, in one of four simultaneous attacks against Western targets; it is impossible to disaggregate the casualty numbers based on the reporting, but, in total, 34 were killed and approximately 60 wounded.
		A suicide bomber detonated a vehicle within the Jedawai housing complex, a residence for foreigners working in Saudi Arabia, in one of four simultaneous attacks against Western targets; it is impossible to disaggregate the casualty numbers based on the reporting, but, in total, 34 were killed and approximately 60 wounded.
		A suicide bomber detonated a vehicle within the Vinnell housing complex, a residence for foreigners working in Saudi Arabia, in one of four simultaneous attacks against Western targets; it is impossible to disaggregate the casualty numbers based on the reporting, but, in total, 34 were killed and approximately 60 wounded.
		An assailant detonated a bomb at the headquarters of the Saudi Maintenance Company in one of four simultaneous attacks against Western targets in Riyadh; no casualties resulted from this attack, though the other three were responsible for killing 34 and wounding approximately 60.
August 5, 2003	Jakarta, Indonesia	A suicide car bomb detonated in front of a Marriott hotel in Jakarta, killing 13 and injuring approximately 149 others.
November 8, 2003	Riyadh, Saudi Arabia	A suicide car bomb detonated outside an upscale villa compound inhabited mainly by Saudis and other Arab nationals, killing at least 17 and injuring more than 120.
November 15, 2003	Istanbul, Turkey	Two suicide truck bombs detonated outside two synagogues in Istanbul, killing least 23 people (plus the two bombers) and injuring more than 300.
November 20, 2003	Istanbul, Turkey	In one of two coordinated attacks, a car bomb was detonated in Istanbul near the British consulate, killing 17; it is impossible to disaggregate the number injured based on the reporting, but, in total, approximately 450 were wounded in the simultaneous attacks.

Table B.1—Continued

Date	Location	Description
November 20, 2003	Istanbul, Turkey	In one of two coordinated attacks, a car bomb was detonated in Istanbul near the HSBC Bank headquarters and the Metro City shopping center, killing 11; it is impossible to disaggregate the number injured based on the reporting, but, in total, approximately 450 were wounded in the simultaneous attacks.
April 27, 2004	Damascus, Syria	Terrorists launched a fierce assault on Damascus' diplomatic district, including detonation of a car bomb close to the British ambassador's residence, wounding one.
May 30, 2004	Khobar, Saudi Arabia	A terrorist shooting rampage and hostage standoff in Saudi Arabia's oil-industry hub killed 22 people, mostly foreigners.
June 8, 2004	Riyadh, Saudi Arabia	Al Qa'ida terrorists gunned down a U.S. defense contractor working for the Vinnell Corp in Saudi Arabia.
June 12, 2004	Riyadh, Saudi Arabia	Al Qa'ida–linked terrorists kidnapped and, days later, executed a U.S. defense contractor working for Lockheed Martin in Riyadh.
		Al Qa'ida–linked terrorists fatally shot a U.S. defense contractor working for the Advanced Electronics Company as he parked his car in front of his villa in Riyadh.
September 15, 2004	Riyadh, Saudi Arabia	Al Qa'ida–linked terrorists fatally shot a British citizen working for the communication firm Marconi in Riyadh.
October 7, 2004	Taba, Egypt	In one of three coordinated attacks, a car bomb was detonated at the Hilton hotel in the Egyptian resort city of Taba, killing approximately 34 and wounding approximately 159.
	Nuweiba, Egypt	In one of three coordinated attacks, a car bomb was detonated at the al-Badiyah campground in Ras al-Shitan, near Nuweiba; casualties suffered during the simultaneous blasts cannot be precisely disaggregated, but approximately five were killed and 12 wounded in both blasts.
	Nuweiba, Egypt	In one of three coordinated attacks, a car bomb was detonated at the Moon Island campgrounds in Ras al-Shitan, near Nuweiba; casualties suffered during the simultaneous blasts cannot be precisely disaggregated, but approximately five were killed and 12 wounded in both blasts.

Table B.1—Continued

Date	Location	Description
October 28, 2004	Islamabad, Pakistan	A bomb was detonated in the lobby of a Marriott hotel in the center of Islamabad, wounding nine.
December 6, 2004	Jeddah, Saudi Arabia	Five terrorists attacked the U.S. consulate in the port city of Jeddah with explosives and gunfire, killing nine and wounding 15.
December 29, 2004	Riyadh, Saudi Arabia	In the second of two coordinated attacks, a suicide car bomb exploded at a center for recruiting emergency troops in Riyadh, killing one and wounding four.
		In the first of two coordinated attacks, a suicide car bomb exploded outside the interior-ministry building in Riyadh, killing two and wounding six.
July 7, 2005	London, UK	In one of four coordinated bombings in central London, seven people were killed on a train at Aldgate station; in total, 52 people were killed and approximately 700 injured in the four attacks.
		In one of four coordinated bombings in central London, 24 people were killed on a train at the King's Cross/ Russell Square station; in total, 52 people were killed and approximately 700 injured in the four attacks.
		In one of four coordinated bombings in central London, 14 people were killed on a No. 30 bus at Tavistock Square; in total, 52 people were killed and approximately 700 injured in the four attacks.
		In one of four coordinated bombings in central London, seven people were killed on a train at Edgware Road; in total, 52 people were killed and approximately 700 injured in the four attacks.
July 23, 2005	Sharm El-Sheikh, Egypt	In one of three coordinated bombings on tourist infrastructure in Sharm El-Sheikh, Egypt, a car bomb was detonated at the Old Market, killing at least 17; in total, 88 were killed and approximately 200 wounded in the three attacks.
		In one of three coordinated bombings on tourist infrastructure in Sharm El-Sheikh, Egypt, a car bomb was detonated at the Ghazala Gardens Hotel, killing at least 45; in total, 88 were killed and approximately 200 wounded in the three attacks.

Table B.1—Continued

Date	Location	Description
July 23, 2005	Sharm El-Sheikh, Egypt	In one of three coordinated bombings on tourist infrastructure in Sharm El-Sheikh, Egypt, a suitcase bomb was detonated near the Moeyenpick Hotel, killing at least six; in total, 88 were killed and approximately 200 wounded in the three attacks.
August 19, 2005	Eliat, Israel	In one of three coordinated attacks, a Katyusha rocket was fired from a warehouse at an airport in a nearby Israeli port, wounding a taxicab driver.
	Aqaba, Jordan	In one of three coordinated attacks, a Katyusha rocket was fired from a warehouse at two U.S. Navy ships—the USS *Ashland* and the USS Kearsarge—docked in the port of Aqaba; the rocket flew over the bow of the USS *Ashland*, landing in a U.S. military-depot warehouse, killing one Jordanian soldier.
	Aqaba, Jordan	In one of three coordinated attacks, a Katyusha rocket was fired from a warehouse at a public hospital in Aqaba, though no injuries or fatalities were reported in this explosion.
November 9, 2005	Amman, Jordan	In one of three coordinated attacks, a suicide bomb was detonated at the Grand Hyatt in Amman; in total, 63 people were killed and more than 100 wounded in the three attacks.
		In one of three coordinated attacks, a suicide bomb was detonated at the Days Inn in Amman; in total, 63 people were killed and more than 100 wounded in the three attacks.
		In one of three coordinated attacks, a suicide bomb was detonated at the Radisson SAS hotel in Amman; in total, 63 people were killed and more than 100 wounded in the three attacks.
December 27, 2005	Kiryat Shmona, Israel	In one of the first direct al Qa'ida–sponsored attacks on Israeli territory, at least three rockets landed in a residential area of Kiryat Shmona (Shemona) in northern Israel, causing property damage but no injuries.
January 10, 2006	Algiers, Algeria	A roadside bomb exploded next to a bus carrying expatriate Western workers, killing one and wounding nine.
February 3, 2006	Kushtia, Bangladesh	Two bombs were targeted the Rashid Agrofood rice mill, owned by a leader of the Bangladesh Nationalist Party (BNP), though no injuries or fatalities were reported.

Table B.1—Continued

Date	Location	Description
February 3, 2006	Kushtia, Bangladesh	Three bombs were targeted at a local BNP leader, identified as Azim, though no injuries or fatalities were reported.
February 24, 2006	Abqaiq, Saudia Arabia	Suicide bombers attacked Saudi Arabia's largest oil-processing facility but were unable to interrupt oil production when guards prevented them from entering the heavily guarded compound; two guards and at least two terrorists were killed in the firefight; four others were injured.
October 19, 2006	Lakhdaria, Algeria	In a coordinated double-bombing, terrorist elements targeted a fuel cistern belonging to the French company Razel and then targeted Algerian police with a second bomb as they responded to the first explosion.
October 30, 2006	Reghaia, Algeria	In a coordinated double-bombing, terrorist elements opened fire on a police station in Reghaia, resulting in significant material damage but no injuries or casualties.
	Dergana, Algeria	In a coordinated double-bombing, terrorist elements detonated a truck bomb at a police station in Dergana, killing three and wounding 24.
January 23, 2007	Gaza, Gaza Strip	A group of nearly 40 assailants, claiming to be members of the al Qa'ida Organization in Palestine, attacked the al-Wahah tourist resort near Gaza, destroying it with explosives. No one was injured.
February 2, 2007	Benchoud, Algeria	A bomb exploded in the municipal stadium of Benchoud during a soccer match, injuring two. The al Qa'ida Organization in the Islamic Maghreb claimed responsibility.
February 6, 2007	Benchoud, Algeria	Members of al Qa'ida Organization in the Islamic Maghreb assassinated the mayor of Benchoud outside his home. The attack occured three days after a similiar attack at a soccer stadium in Benchoud.
February 12, 2007	Si-Mustafa, Algeria	In one of seven coordinated bombings targeting Algerian police forces in the countryside east of Algiers, a bomb was detonated by the al Qa'ida affiliate known as al Qa'ida Organization in the Islamic Maghreb; in total, six people were killed and 30 wounded in the coordinated attacks.

Table B.1—Continued

Date	Location	Description
February 12, 2007	Draa Benkheda, Algeria Meklaa, Algeria Illoula Oumalou, Algeria Meklaa, Algeria Souk El Had, Algeria Souk El Had, Algeria	In one of seven coordinated bombings targeting Algerian police forces in the countryside east of Algiers, a bomb was detonated by the al Qa'ida affiliate known as al Qa'ida Organization in the Islamic Maghreb; in total, six people were killed and 30 wounded in the coordinated attacks.
March 3, 2007	Ain Delfa, Algeria	Al Qa'ida Organization in the Islamic Maghreb planted a bomb that later exploded as a convoy carrying Russian workers passed, killing seven and wounding five.
March 16, 2007	Bayt Hanun, Gaza Strip	Militants belonging to the affiliate al Qa'ida in Palestine fired 11 bullets at the car of the director of the United Nations Relief and Works Agency in Gaza, who escaped the assassination attempt; no one was injured in the attack.
March 28, 2007	Boumerdes, Algeria	Al Qa'ida Organization in the Islamic Maghreb remotely detonated a bomb in downtown Boumerdes. The bomb's intended target was the police chief, but his convoy passed before the bomb detonated, causing only limited damage to surrounding buildings.
April 11, 2007	Algiers, Algeria	Al Qa'ida Organization in the Islamic Maghreb detonated three suicide car bombs near a government building in an apparent assassination attempt against the prime minister, killing 23 and wounding 162 others. The bombings were coordinated to coincide with a second attack against an Algiers police station.
	Algiers, Algeria	Militants from al Qa'ida Organization in the Islamic Maghreb detonated a bomb at a police station in Algiers, killing eight and wounding 50. The bombing was coordinated to coincide with a second attack on the office of the prime minister.

Table B.1—Continued

Date	Location	Description
April 25, 2007	Boumerdes, Algeria	Two militants from al Qa'ida Organization in the Islamic Maghreb killed a member of a citizen militia in a market in Boumerdes.
May 13, 2007	Constantine, Algeria	One police officer was killed and two other people injured when militants belonging to al Qa'ida Organization in the Islamic Maghreb detonated a homemade bomb at a police checkpoint; the blast came days before a national election.
May 15, 2007	Timidaouen, Algeria	Militants from al Qa'ida Organization in the Islamic Maghreb damaged the gas mains of the town of Timidaouen with a homemade bomb, though the explosion caused no casualities.
July 11, 2007	Lakhdaria, Algeria	A suicide bomber belonging to al Qa'ida Organization in the Islamic Maghreb drove a refrigerated truck filled with explosives deep into a military barracks, killing 10 and wounding 23.
July 20, 2007	Souk El Had, Algeria	Nine railroad cars were derailed when al Qa'ida Organization in the Islamic Maghreb launched an attack against a fuel freight train; the attack damaged roughly 30 meters of railroad track and injured one soldier.
September 6, 2007	Batna, Algeria	In an assassination attempt against President Abdelaziz Bouteflika, an 18-year-old member of al Qa'ida Organization in the Islamic Maghreb detonated a bomb, killing 15 residents of the town of Batna and injuring more than 75.
	Dellys, Algeria	Only 48 hours after launching a similar attack, a 15-year-old member of al Qa'ida Organization in the Islamic Maghreb detonated a truck loaded with explosives, killing at least 30 people. All of the victims were coast-guard troops engaged in a flag-raising ceremony.
September 12, 2007	Si-Mustafa, Algeria	In an attempt to create an electrical blackout, al Qa'ida Organization in the Islamic Maghreb destroyed three electrical-energy pylons with explosives. The attack was only successful in a limited sense, however, as authorities quickly turned to other electrical-energy resources.
September 21, 2007	Lakhdaria, Algeria	Nine people were wounded when a suicide bomber belonging to al Qa'ida Organization in the Islamic Maghreb drove an explosive-laden vehicle into a truck convoy about 100 kilometers outside of Algiers. The attack, which appeared to target foreign employees of a construction firm, came only hours after al Qa'ida issued a call for attacks against French targets.

Table B.1—Continued

Date	Location	Description
December 11, 2007	Algiers, Algeria	In one of two coordinated suicide bombings, a member of al Qa'ida Organization in the Islamic Maghreb detonated an explosive-laden vehicle outside the United Nations office for the High Commissioner for Refugees and the UN Development Program building; in total, approximately 67 people were killed and 177 wounded in both attacks, though conflicting news reports exist on these precise figures.
		In one of two coordinated suicide bombings, a member of al Qa'ida Organization in the Islamic Maghreb detonated an explosive-laden vehicle outside the Algerian constitutional court, destroying a nearby bus packed with university students; in total, approximately 67 people were killed and 177 wounded in both attacks, though conflicting news reports exist on these precise figures.
December 27, 2007	Rawalpindi, Pakistan	Members of al Qa'ida and allies of Pakistani tribal leader Baitullah Mehsud fired two shots at former prime minister Benazir Bhutto and then detonated a suicide bomb after a political rally at Liaquat National Bagh, killing 24 people and wounding approximately 50 others.

APPENDIX C

Regression Analysis

Using regression analysis to try to explain which terrorist groups lasted how long or how they ended entails creating a dependent variable (e.g., lifetime) and seeing whether it is correlated with one or more independent variables (e.g., its primary classification). Our analysis refuted many hypotheses and confirmed none. The only variable that correlated with terrorist duration (explained in this appendix) was the group's peak size. Essentially, the larger terrorist groups were those that lasted longer and were more likely either to achieve their outcomes or to close up shop consistently with their goals (e.g., politics). But causality could work the other way: Longevity, in part, may explain size.

In performing this task, we created a single "success" measure for terrorist groups that had to include how they ended (e.g., in victory or not), how long they lasted, and (because many such groups are of recent formation) whether they were still active. Our measure was built on a point system, composed of three factors:

- *Longevity:* The points awarded equal the square root of the number of years between the group's founding and its ending, or 2006, if they were alive as of that date.
- *Survival:* If the group was alive, the points awarded for the longevity measure were doubled—but in no case was allowed to exceed 10 points.
- *Ending:* We added 10 points if the group ended because of victory and two points if it ended because of politics or splintering (conversely, a group that ended because of police or military pressure received no extra points).

Thus, a group that did not live long and was suppressed by authorities received few points. One that survived for decades and is still around or, for example, entered the political process, received more points.

We examined several independent variables:

- *Host-country income:* ranging from 3 (high income as measured by the World Bank) to 2 (upper middle), 1 (lower middle), and 0 (lower income)
- *Host-country freedom:* ranging from 2 (free) to 1 (partially free) and 0 (not free)
- *Peak size:* ranging from 3 (more than 10,000 members) to 2 (1,000 to 9,999 members), 1 (100 to 999 members), and 0 (99 or fewer members)
- *Is left-wing:* 1 if a group can be considered left-wing, 0 otherwise. Similarly for *is nationalist* and *is religious* (for right-wing groups, the value of each of those three variables is zero).
- *Breadth:* Following Figure 2.2 in Chapter Two, a terrorist group's goals were graded according to whether they were narrow or broad—that is, whether it would be easier or harder to accommodate. The status-quo goal was considered the narrowest, and this was scored as 0. This was followed by the policy-change goal (scored as 1), the territorial-change goal (2), the regime-change goal (3), the empire goal (4), and the social-revolution goal (5).

The best single-factor correlation was with the peak size, which alone explained 32 percent of the variance in outcome scores (i.e., $R^2 = 0.32$). No other single variable alone explained more than 4 percent of the variance. Remarkably, the addition of all other explanatory variables together improved the explanatory power only from 32 to 34 percent, or hardly at all.[1] For the sake of reference, Table C.1 presents the results with the peak-size variable only. And Table C.2 presents the results of all explanatory variables working together.

---

[1] The adjusted R^2 is 33 percent. Adjustment compensates for the truism that one can explain away all variance by throwing in enough explanatory factors.

Table C.1
Peak-Size Variable

Term	Coefficient	Standard Error	p	95% Confidence Interval (CI) of Coefficient	
				Low	High
Intercept	3.2960	0.1318	<0.0001	3.0373	3.5548
Slope	2.0978	0.1210	<0.0001	1.8601	2.3354

NOTE: R^2 and adjusted R^2 are 0.32. Standard error is 2.6809.

Table C.2
All Explanatory Variables

Term	Coefficient	Standard Error	p	95% CI of Coefficient	
				Low	High
Intercept	2.3689	0.5786	<0.0001	1.2326	3.5052
Income	0.0100	0.1168	0.9320	−0.2193	0.2393
Freedom	0.2939	0.1589	0.0647	−0.0180	0.6058
Peak size	2.1004	0.1288	<0.0001	1.8474	2.3534
Is left-wing	0.6129	0.4550	0.1785	−0.2806	1.5064
Is nationalist	1.2417	0.4537	0.0064	0.3508	2.1327
Is religious	1.2648	0.4955	0.0109	0.2918	2.2377
Breadth	−0.1333	0.1173	0.2562	−0.3636	0.0970

NOTE: R^2 is 0.34. Adjusted R^2 is 0.33. Standard error is 2.6574.

Lest it appear that *is nationalist* or *is religious* has more than modest explanatory power, note the results with just *peak size* and *is religious* alone (Table C.3). And Table C.4 presents the results of *peak size* and *is nationalist* alone.

Table C.3
Peak-Size and Is-Religious Variables

Term	Coefficient	Standard Error	p	95% CI of Coefficient	
				Low	High
Intercept	3.2612	0.1419	<0.0001	2.9825	3.5399
Peak size	2.0947	0.1212	<0.0001	1.8568	2.3327
Is religious	0.1691	0.2555	0.5083	–0.3327	0.6709

NOTE: R^2 and adjusted R^2 are 0.32. Standard error is 2.6821.

Table C.4
Peak-Size and Is-Nationalist Variables

Term	Coefficient	Standard Error	p	95% CI of Coefficient	
				Low	High
Intercept	3.0983	0.1497	<0.0001	2.8043	3.3922
Peak size	2.0739	0.1208	<0.0001	1.8367	2.3110
Is nationalist	0.6011	0.2196	0.0064	0.1700	1.0323

NOTE: R^2 is 0.33. Adjusted R^2 is 0.32. Standard error is 2.6676.

References

9/11 Commission—*see* National Commission on Terrorist Attacks upon the United States.

Abrahms, Max, "Why Terrorism Does Not Work," *International Security*, Vol. 31, No. 2, Fall 2006, pp. 42–78.

Adams, Gordon, *Budgeting for Iraq and the GWOT: Testimony, Committee on the Budget, United States Senate*, February 6, 2007.

Al Jazeera, interview with Mullah Dadullah, July 2005.

"Al Jazeera Reveals New al Qaeda Leader," *Washington Times*, May 25, 2007, p. 17.

Al-Ansary, Khalid, and Ali Adeeb, "Most Tribes in Anbar Agree to Unite Against Insurgents," *New York Times*, September 18, 2006, p. A12.

Al-Zawahiri, Ayman, letter to Abu Musab al-Zarqawi, July 9, 2005.

———, "Knights Under the Prophet's Banner," in John Calvert, ed., *Islamism: A Documentary and Reference Guide*, Westport, Conn.: Greenwood Press, 2008.

Alden, Chris, *Mozambique and the Construction of the New African State: From Negotiations to Nation Building*, New York: Palgrave, 2001.

Americas Watch Committee, *The Killings in Colombia*, Washington, D.C., 1989.

———, *El Salvador's Decade of Terror: Human Rights Since the Assassination of Archbishop Romero*, New Haven, Conn.: Yale University Press, 1991.

Asahara, Shōkō, *Disaster Approaches the Land of the Rising Sun: Shoko Asahara's Apocalyptic Predictions*, Shizuoka, Japan: Aum Publishing Co., 1995.

"Aum Admits Matsumoto May Be Linked to Crimes," *Daily Yomiuri*, January 19, 2000, p. 1.

Beblawi, Hazem, and Giacomo Luciani, eds., *The Rentier State*, London and New York: Croom Helm, 1987.

Belasco, Amy, *The Cost of Iraq, Afghanistan, and Other Global War on Terror Operations Since 9/11*, Washington, D.C.: Congressional Research Service, Library of Congress, 2007.

Benjamin, Daniel, and Steven Simon, *The Age of Sacred Terror*, New York: Random House, 2002.

Peter Bergen, "Al Qaeda Status," written testimony submitted to the U.S. House of Representatives Permanent Select Committee on Intelligence, April 9, 2008.

Berntsen, Gary, *First Directive*, unpublished manuscript, August 2007.

Berntsen, Gary, and Ralph Pezzullo, *Jawbreaker: The Attack on Bin Laden and Al Qaeda*, New York: Crown Publishers, 2005.

Biddle, Stephen D., *Afghanistan and the Future of Warfare: Implications for Army and Defense Policy*, Carlisle, Pa.: Strategic Studies Institute, U.S. Army War College, November 2002.

Bin Laden, Osama, "Declaration of War Against the Americans Occupying the Land of the Two Holy Places," August 23, 1996.

———, "World Islamic Front for Jihad Against Jews and Crusaders: Initial 'Fatwa' Statement," *al-Quds al-Arabi*, February 23, 1998, p. 3.

Bloom, Mia, *Dying to Kill: The Allure of Suicide Terror*, New York: Columbia University Press, 2005.

Boot, Max, "Can Petraeus Pull It Off?" *Weekly Standard*, April 30, 2007, p. 24.

Boutros-Ghali, Boutros, "Report of the Secretary-General on the United Nations Observer Mission in El Salvador (Onusal)," New York: United Nations, S/25006, December 23, 1992.

———, *From Madness to Hope: The 12-Year War in El Salvador*, New York: United Nations, 1993a.

———, "Letter Dated 93/01/07 from the Secretary-General Addressed to the President of the Security Council," New York: United Nations, S/25078, January 9, 1993b.

———, "Report of rhe Secretary-General on the United Nations Observer Mission in El Salvador," New York: United Nations, S/1994/179, February 6, 1994a.

———, "Report of the Secretary-General on the United Nations Observer Mission in El Salvador," New York: United Nations, S/1994/304, March 16, 1994b.

———, "Report of the Secretary-General on the United Nations Observer Mission in El Salvador," New York: United Nations, S/1994/375, March 31, 1994c.

———, "Report of the Secretary-General on the United Nations Observer Mission in El Salvador," New York: United Nations, S/1995/220, March 24, 2005.

Bracket, D. W., *Holy Terror: Armageddon in Tokyo*, New York: Weatherhill, 1996.

Brands, H. W., *The Devil We Knew: Americans and the Cold War*, New York: Oxford University Press, 1993.

Bremer, Paul, director of reconstruction and humanitarian assistance, "Message for SecDef," email to Jaymie Durnan, special assistant to the deputy defense secretary, June 30, 2003.

Burns, John F., "Showcase and Chimera in the Desert," *New York Times*, July 8, 2007, p. A1.

Burns, John F., and Alissa J. Rubin, "U.S. Arming Sunnis in Iraq to Battle Old Qaeda Allies," *New York Times*, June 11, 2007, p. A1.

Bush, George W., *State of the Union*, Washington, D.C.: White House, 2002.

———, *The National Security Strategy of the United States of America*, Washington, D.C.: White House, 2006.

Byman, Daniel, *The Five Front War: The Better Way to Fight Global Jihad*, Hoboken, N.J.: John Wiley and Sons, 2008.

Byrne, Hugh, *El Salvador's Civil War: A Study of Revolution*, Boulder, Colo.: Lynne Rienner Publishers, 1996.

Cabarrús, Carlos Rafael, *Génesis de una Revolución: Análisis del Surgimiento y Desarrollo de la Organización Campesina en El Salvador*, Mexico: Centro de Investigaciones y Estudios Superiores en Antropología Social, 1983.

Call, Charles T., "Institutional Learning Within ICITAP," in Robert B. Oakley, Michael J. Dziedzic, and Eliot M. Goldberg, eds., *Policing the New World Disorder: Peace Operations and Public Security*, Washington, D.C.: National Defense University Press, 1998, pp. 315–364.

———, "Assessing El Salvador's Transition from Civil War to Peace," in Stephen John Stedman, Donald S. Rothchild, and Elizabeth M. Cousens, *Ending Civil Wars: The Implementation of Peace Agreements*, Boulder, Colo.: Lynne Rienner, 2002, pp. 383–420.

———, "Democratisation, War and State-Building: Constructing the Rule of Law in El Salvador," *Journal of Latin American Studies*, Vol. 35, No. 4, November 2003, p. 831.

Callwell, C. E., *Small Wars: Their Principles and Practices*, 3rd ed., Lincoln, Neb.: University of Nebraska Press, 1996.

Cardenal, Rodolfo, *Historia de una Esperanza: Vida de Rutilio Grande*, San Salvador, El Salvador: UCA Editores, 1985.

Castaneda, Ricardo G., ambassador and permanent representative, "Letter Dated 91/09/26 from the Permanent Representative of El Salvador to the United Nations Addressed to the Secretary-General," New York: United Nations, A/46/502-S/23082, September 26, 1991a.

———, "Letter Dated 91/10/08 from the Permanent Representative of El Salvador to the United Nations Addressed to the Secretary-General," New York: United Nations, A/46/551-S/23128, October 9, 1991b.

Celeski, Joseph D., *Operationalizing COIN*, Hurlburt Field, Fla.: JSOU Press, report 05-2, 2005.

Cheney, Dick, "Vice Presidents Remarks at a Rally for the Troops," USS *John C. Stennis*, May 11, 2007.

Childress, Michael, *The Effectiveness of U.S. Training Efforts in Internal Defense and Development: The Cases of El Salvador and Honduras*, Santa Monica, Calif.: RAND Corporation, MR-250-USDP, 1995. As of March 7, 2008:
http://www.rand.org/pubs/monograph_reports/MR250/

Childress, Sarah, "Retaking Ramadi: All the Sheik's Men: U.S. Commanders Are Hoping Tribal Levies Can Help Fill the Ranks of Anbar's Police and Tackle Al Qaeda," *Newsweek*, December 18, 2006, p. 43.

Clutterbuck, Lindsay, "Law Enforcement," in Audrey Kurth Cronin and James M. Ludes, eds., *Attacking Terrorism: Elements of a Grand Strategy*, Washington, D.C.: Georgetown University Press, 2004, pp. 140–161.

Collier, David, "New Perspectives on the Comparative Method," in Dankwart A. Rustow and Kenneth Paul Erickson, eds., *Comparative Political Dynamics: Global Research Perspectives*, New York: HarperCollins, 1991, pp. 7–31.

Connaughton, Richard, "Operation 'Barass,'" *Small Wars and Insurgencies*, Vol. 12, No. 2, Summer 2001, pp. 110–119.

Constable, Pamela, "Gates Visits Kabul, Cites Rise in Cross-Border Attacks," *Washington Post*, January 17, 2007, p. A10.

Cortright, David, and George A. Lopez, *The Sanctions Decade: Assessing UN Strategies in the 1990s*, Boulder, Colo.: Lynne Rienner Publishers, 2000.

Crenshaw, Martha, "The Causes of Terrorism," in Charles W. Kegley, ed., *International Terrorism: Characteristics, Causes, Controls*, New York: St. Martin's, 1990, pp. 92–105.

———, "Why Violence Is Rejected or Renounced: A Case Study of Oppositional Terrorism," in Thomas Gregor, ed., *A Natural History of Peace*, Nashville, Tenn.: Vanderbilt University Press, 1996, pp. 249–272.

Cronin, Audrey Kurth, "Behind the Curve: Globalization and International Terrorism," *International Security*, Vol. 27, No. 3, Winter 2002–2003, pp. 30–58.

———, "How al-Qaida Ends: The Decline and Demise of Terrorist Groups," *International Security*, Vol. 31, No. 1, Summer 2006, pp. 7–48.

———, *Ending Terrorism: Lessons for Defeating al-Qaeda*, Abingdon, Oxon: Routledge for the International Institute for Strategic Studies, 2008.

Cronin, Audrey Kurth, and James M. Ludes, eds., *Attacking Terrorism: Elements of a Grand Strategy*, Washington, D.C.: Georgetown University Press, 2004.

Dagher, Sam, "Sunni Muslim Sheikhs Join US in Fighting al Qaeda," *Christian Science Monitor*, May 3, 2007, p. 1.

Del Castillo, Graciana, "The Arms-for-Land Deal in El Salvador," in Michael W. Doyle, Ian Johnstone, and Robert C. Orr, eds., *Keeping the Peace: Multidimensional UN Operations in Cambodia and El Salvador*, New York: Cambridge University Press, 1997, pp. 342–366.

Dershowitz, Alan M., *Why Terrorism Works: Understanding the Threat, Responding to the Challenge*, New Haven, Conn.: Yale University Press, 2002.

DNI—*see* Office of the Director of National Intelligence.

Dobbins, James, Seth G. Jones, Keith Crane, and Beth Cole DeGrasse, *The Beginner's Guide to Nation-Building*, Santa Monica, Calif.: RAND Corporation, MG-557-SRF, 2007. As of March 13, 2008:
http://www.rand.org/pubs/monographs/MG557/

Dobbins, James, Seth G. Jones, Keith Crane, Andrew Rathmell, Brett Steele, Richard Teltschik, and Anga R. Timilsina, *The UN's Role in Nation-Building: From the Congo to Iraq*, Santa Monica, Calif.: RAND Corporation, MG-304-RC, 2005. As of March 13, 2008:
http://www.rand.org/pubs/monographs/MG304/

Dobbins, James, John G. McGinn, Keith Crane, Seth G. Jones, Rollie Lal, Andrew Rathmell, Rachel M. Swanger, and Anga R. Timilsina, *America's Role in Nation-Building: From Germany to Iraq*, Santa Monica, Calif.: RAND Corporation, MR-1753-RC, 2003. As of March 13, 2008:
http://www.rand.org/pubs/monograph_reports/MR1753/

DoD—*see* U.S. Department of Defense.

DoS—*see* U.S. Department of State.

Doyle, Michael W., Ian Johnstone, and Robert C. Orr, eds., *Keeping the Peace: Multidimensional UN Operations in Cambodia and El Salvador*, New York: Cambridge University Press, 1997.

Doyle, Michael W., and Nicholas Sambanis, *Making War and Building Peace: United Nations Peace Operations*, Princeton, N.J.: Princeton University Press, 2006.

Edelstein, David, "Occupational Hazards: Why Military Occupations Succeed or Fail," *International Security*, Vol. 29, No. 1, Summer 2004, pp. 49–91.

Eisenstadt, Michael, and Jeffrey White, *Assessing Iraq's Sunni Arab Insurgency*, Washington, D.C.: The Washington Institute for Near East Policy, Policy Focus 50, December 2005.

Embassy of El Salvador, "The Peace Accords," undated Web page. As of April 14, 2008:
http://www.elsalvador.org/embajadas/eeuu/home.nsf/politics

Evans, Jonathan, "Address to the Society of Editors by the Director General of the Security Service, Jonathan Evans," Society of Editors' "A Matter of Trust" conference, Radisson Edwardian Hotel, Manchester, UK, November 5, 2007.

Fadel, Leila, "Iraq: Old Allegiances Loom Large as U.S. Trains Iraqi Forces," *Miami Herald*, June 17, 2007, p. A24.

Fearon, James D., and David D. Laitin, "Ethnicity, Insurgency, and Civil War," *American Political Science Review*, Vol. 97, No. 1, February 2003, pp. 75–90.

Feith, Douglas J., *War and Decision: Inside the Pentagon at the Dawn of the War on Terror*, New York: HarperCollins Publishers, 2008.

Freedberg, Sydney J. Jr., "The New Iraqi Way of War," *National Journal*, Vol. 39, No. 23, June 9, 2007, pp. 36–43.

Freedom House, *Freedom in the World 2007: The Annual Survey of Political Rights and Civil Liberties*, New York: Freedom House, 2007.

Fromkin, David, "The Strategy of Terrorism," *Foreign Affairs*, Vol. 53, No. 4, July 1975, pp. 683–698.

Gaddis, John Lewis, *Strategies of Containment: A Critical Appraisal of Postwar American National Security Policy*, New York: Oxford University Press, 1982.

Galula, David, *Counterinsurgency Warfare: Theory and Practice*, New York: Praeger, 1964.

———, *Counterinsurgency Warfare: Theory and Practice*, St. Petersburg, Fla.: Hailer Publishing, 2005.

George, Alexander L., "Case Studies and Theory Development: The Method of Structured, Focused Comparison," in Paul Gordon Lauren, ed., *Diplomacy: New Approaches in History, Theory, and Policy*, New York: Free Press, 1979, pp. 43–68.

George, Alexander L., and Timothy J. McKeown, "Case Studies and Theories of Organizational Decision Making," in Robert F. Coulam and Richard A. Smith, *Advances in Information Processing in Organizations: A Research Annual*, Vol. 2, Greenwich, Conn.: JAI Press, 1985, pp. 21–58.

Gompert, David C., and James Dobbins, "A Far Too Costly Pentagon," United Press International, February 27, 2006.

Gompert, David C., John Gordon IV, Adam Grissom, David R. Frelinger, Seth G. Jones, Martin C. Libicki, Edward O'Connell, Brooke K. Stearns, and Robert E. Hunter, *War by Other Means: Building Complete and Balanced Capabilities for Counterinsurgency (RAND Counterinsurgency Study: Final Report)*, Santa Monica, Calif.: RAND Corporation, MG-595/2-OSD, 2008. As of March 11, 2008: http://www.rand.org/pubs/monographs/MG595.2/

Gordon, Michael R., "Grim Report Out of Anbar Is Disputed by General," *New York Times*, September 13, 2006, p. A15.

———, "G.I.'s Forge Sunni Tie in Bid to Squeeze Militants," *New York Times*, July 6, 2007, p. A1.

Great Britain Northern Ireland Office, *The Belfast Agreement: An Agreement Reached at the Multi-Party Talks on Northern Ireland*, London: Stationery Office, 1998.

Guide to the Analysis of Insurgency, 1986.

Guidère, Mathieu, and Peter Harling, "Iraq's Resistance Evolves," *Le Monde Diplomatique*, May 2006.

Gunaratna, Rohan, *Inside Al Qaeda: Global Network of Terror*, New York: Columbia University Press, 2002.

———, "Combating the Al-Qaeda Associated Groups," in Doron Zimmermann and Andreas Wenger, eds., *How States Fight Terrorism: Policy Dynamics in the West*, Boulder, Colo.: Lynne Rienner Publishers, 2007, pp. 175–202.

Gurr, Ted Robert, "Terrorism in Democracies: Its Social and Political Bases," in Walter Reich, ed., *Origins of Terrorism: Psychologies, Ideologies, Theologies, States of Mind*, Washington, D.C.: Woodrow Wilson International Center for Scholars, 1990, pp. 86–102.

Harub, Khalid, *Hamas: Political Thought and Practice*, Washington, D.C.: Institute for Palestine Studies, 2000.

Hashim, Ahmed, *Insurgency and Counter-Insurgency in Iraq*, Ithaca, N.Y.: Cornell University Press, 2006.

Hastert, Paul L., "Operation Anaconda: Perception Meets Reality in the Hills of Afghanistan," *Studies in Conflict and Terrorism*, Vol. 28, No. 1, January–February 2005.

Hatfield, Mark O., James Leach, and George Miller, *Bankrolling Failure: United States Policy in El Salvador and the Urgent Need for Reform: A Report to the Arms Control and Foreign Policy Caucus*, Washington, D.C., 1987.

Hazan, D., *Sunni Jihad Groups Rise Up Against Al-Qaeda in Iraq*, Washington, D.C.: Middle East Media Research Institute, Inquiry and Analysis Series 336, March 22, 2007.

Hewitt, Christopher, *Understanding Terrorism in America: From the Klan to al Qaeda*, New York: Routledge, 2002.

Heymann, Philip B., *Terrorism and America: A Commonsense Strategy for a Democratic Society*, Cambridge, Mass.: MIT Press, 1998.

———, "Dealing with Terrorism: An Overview," *International Security*, Vol. 26, No. 3, Winter 2001, pp. 24–38.

Hideaki, Mizukoshi, "Terrorists, Terrorism, and Japan's Counter-Terrorism Policy," *Gaiko Forum*, Vol. 3, No. 2, Summer 2003, pp. 53–63.

Hoffman, Bruce, *Inside Terrorism*, 2nd ed., New York: Columbia University Press, 2006.

———, "Challenges for the U.S. Special Operations Command Posed by the Global Terrorist Threat: Al Qaeda on the Run or on the March?" written testimony submitted to the U.S. House of Representatives Committee on Armed Services Subcommittee on Terrorism, Unconventional Threats, and Capabilities, February 14, 2007.

House of Commons, *Report of the Official Account of the Bombings in London on 7th July 2005*, London: The Stationery Office, HC 1087, 2006.

Hoyt, Timothy D., "Military Force," in Audrey Kurth Cronin and James M. Ludes, eds., *Attacking Terrorism: Elements of a Grand Strategy*, Washington, D.C.: Georgetown University Press, 2004, pp. 162–185.

Huggins, Martha Knisely, *Political Policing: The United States and Latin America*, Durham, N.C.: Duke University Press, 1998.

Hughes, Christopher W., "The Reaction of the Police and Security Authorities to Aum Shinrikyo," in Robert Kisala and Mark Mullins, eds., *Religion and Social Crisis in Japan: Understanding Japanese Society Through the Aum Affair*, Basingstoke, Hampshire, UK, and New York: Palgrave, 2001, p. 65.

Huntington, Samuel P., "The Age of Muslim Wars," *Newsweek*, December 17, 2001, p. 48.

ICG—*see* International Crisis Group.

IISS—*see* International Institute for Strategic Studies.

International Crisis Group, *In Their Own Words: Reading the Iraqi Insurgency*, Amman and Brussels, 2006.

International Institute for Strategic Studies, *The Military Balance, 1991–1992*, London: Brassey's, 1991.

Interview with Afghan government officials, Kabul, Afghanistan, August 2006.

Interview with member of l'Unité de Coordination de la Lutte Antiterroriste by Seth G. Jones, Monterey, Calif., June 27, 2007.

Interview with members of l'Unité de Coordination de la Lutte Antiterroriste by Seth G. Jones, Monterey, Calif., January 22, 2008.

Interview with UK intelligence official by Seth G. Jones, Washington, D.C., January 23, 2008.

"Iraq's Sunni Armed Groups Reportedly Planning Alliance Against Al-Qa'ida," *Al-Hayay* (London), April 11, 2007.

Jackson, Brian A., "Groups, Networks, or Movements: A Command-and-Control–Driven Approach to Classifying Terrorist Organizations and Its Application to Al Qaeda," *Studies in Conflict and Terrorism*, Vol. 29, No. 3, May 2006, pp. 241–262.

Jaffe, Greg, "How Courting Sheiks Slowed Violence in Iraq," *Wall Street Journal*, August 8, 2007, p. A1.

Jalali, Ali A., "The Future of Afghanistan," *Parameters*, Vol. 36, No. 1, Spring 2006, pp. 4–19.

Jervis, Rick, "Police in Iraq See Jump in Recruits," *USA Today*, January 15, 2007, p. 1.

Johnson, Chris, "Military Officers Criticize Media Coverage of Battlefront in Iraq," *Inside the Navy*, February 5, 2007, p. 5.

Johnson, Stuart, Martin Libicki, and Gregory F. Treverton, eds., *New Challenges, New Tools for Defense Decisionmaking*, Santa Monica, Calif.: RAND Corporation, MR-1576-RC, 2003. As of March 13, 2008:
http://www.rand.org/pubs/monograph_reports/MR1576/

Johnstone, Ian, "Rights and Reconciliation in El Salvador," in Michael W. Doyle, Ian Johnstone, and Robert C. Orr, eds., *Keeping the Peace: Multidimensional UN Operations in Cambodia and El Salvador*, New York: Cambridge University Press, 1997, pp. 312–341.

Joint Group for the Investigation of Politically Motivated Illegal Armed Groups in El Salvador, and United Nations Security Council, *Report of the Joint Group for the Investigation of Politically Motivated Illegal Armed Groups in El Salvador: San Salvador, 28 July 1994*, New York: United Nations Security Council, S/1994/989, October 22, 1994.

Jones, Seth G., *The Rise of European Security Cooperation*, Cambridge and New York: Cambridge University Press, 2007a.

———, "Fighting Networked Terror Groups: Lessons from Israel," *Studies in Conflict and Terrorism*, Vol. 30, No. 4, April 2007b, pp. 281–302.

———, *Counterinsurgency in Afghanistan: RAND Counterinsurgency Study, Volume 4*, Santa Monica, Calif.: RAND Corporation, MG-595-OSD, 2008. As of June 27, 2008:
http://www.rand.org/pubs/monographs/MG595/

Jones, Seth G., Olga Oliker, Peter Chalk, C. Christine Fair, Rollie Lal, and James Dobbins, *Securing Tyrants or Fostering Reform? U.S. Internal Security Assistance to Repressive and Transitioning Regimes*, Santa Monica, Calif.: RAND Corporation, MG-550-OSI, 2006. As of March 7, 2008:
http://www.rand.org/pubs/monographs/MG550/

Jones, Seth G., Jeremy M. Wilson, Andrew Rathmell, and K. Jack Riley, *Establishing Law and Order After Conflict*, Santa Monica, Calif.: RAND Corporation, MG-374-RC, 2005. As of March 13, 2008:
http://www.rand.org/pubs/monographs/MG374/

Juergensmeyer, Mark, *Terror in the Mind of God: The Global Rise of Religious Violence*, Berkeley, Calif.: University of California Press, 2000.

Kalyvas, Stathis N., *The Logic of Violence in Civil War*, Cambridge and New York: Cambridge University Press, 2006.

Kaplan, David E., and Andrew Marshall, *The Cult at the End of the World: The Terrifying Story of the Aum Doomsday Cult, from the Subways of Tokyo to the Nuclear Arsenals of Russia*, New York: Crown Publishers, 1996.

Karl, Terry Lynn, "El Salvador at the Crossroads: Negotiations or Total War," *World Policy Journal*, Vol. 6, No. 2, April 1989, pp. 321–355.

Karzai, Hekmat, *Afghanistan and the Globalisation of Terrorist Tactics*, Singapore: Institute of Defence and Strategic Studies, January 2006.

Karzei, Hekmat, and Seth G. Jones, "How to Curb Rising Suicide Terrorism in Afghanistan," *Christian Science Monitor*, July 18, 2006.

Katzenstein, Peter J., "Same War: Different Views: Germany, Japan, and Counterterrorism," *International Organization*, Vol. 57, No. 4, Autumn 2003, pp. 731–760.

Kennedy, John F., "Special Message to the Congress on the Defense Budget," March 28, 1961, in John F. Kennedy, *Public Papers of the Presidents of the United States: John F. Kennedy; Containing the Public Messages, Speeches, and Statements of the President, January 20 to December 31, 1961*, Washington, D.C.: U.S. Government Printing Office, 1962, p. 236.

———, "Development of U.S. and Indigenous Police, Paramilitary and Military Resources," national security action memorandum 162, June 19, 1962.

Khalil, Lydia, "The Islamic State of Iraq Launches Plan of Nobility," *Terrorism Focus*, Vol. 4, No. 7, March 27, 2007.

Klein, Hans-Joachim, *La Mort Mercenaire: Témoignage d'un Ancient Terroriste Oust-Allemand*, Paris: Seuil, 1980.

Klein, Joe, "Saddam's Revenge," *Time*, September 26, 2005, pp. 44–51.

Kosal, Margaret, "Terrorism Targeting Industrial Chemical Facilities: Strategic Motivations and the Implications for U.S. Security," *Studies in Conflict and Terrorism*, Vol. 29, No. 7, October 2006, pp. 719–751.

Kosiak, Steven M., *The Global War on Terror (GWOT): Costs, Cost Growth and Estimating Funding Requirements*, testimony before the U.S. Senate Committee on Budget, February 6, 2007.

Kramer, Mark, "The Perils of Counterinsurgency: Russia's War in Chechnya," *International Security*, Vol. 29, No. 3, Winter 2004–2005, pp. 5–62.

Kraul, Chris, "In Ramadi, a Ragtag Solution with Real Results," *Los Angeles Times*, May 7, 2007, p. A6.

Kull, Steven, Clay Ramsay, Stephen Weber, Evan Lewis, Ebrahim Mohseni, Mary Speck, Melanie Ciolek, and Melinda Brouwer, *Muslim Public Opinion on US Policy, Attacks on Civilians and al Qaeda*, College Park, Md.: WorldPublicOpinion.org, Program on International Policy Attitudes, University of Maryland, 2007.

Kydd, Andrew, and Barbara F. Walter, "Sabotaging the Peace: The Politics of Extremist Violence," *International Organization*, Vol. 56, No. 2, Spring 2002, pp. 263–296.

———, "The Strategies of Terrorism," *International Security*, Vol. 31, No. 1, Summer 2006, pp. 49–79.

Lake, David A., "Rational Extremism: Understanding Terrorism in the Twenty-First Century," *Dialogue IO*, Vol. 1, No. 1, January 2002, pp. 15–29.

Levine, Mark, "Peacemaking in El Salvador," in Michael W. Doyle, Ian Johnstone, and Robert C. Orr, eds., *Keeping the Peace: Multidimensional UN Operations in Cambodia and El Salvador*, New York: Cambridge University Press, 1997, pp. 227–255.

Lewis, Bernard, *The Crisis of Islam: Holy War and Unholy Terror*, New York: Modern Library, 2003.

Libicki, Martin C., *The Mesh and the Net: Speculations on Armed Conflict in a Time of Free Silicon*, Washington, D.C.: Institute for National Strategic Studies, National Defense University, 1994.

———, *Information Technology Standards: Quest for the Common Byte*, Boston: Digital Press, 1995a.

———, *What Is Information Warfare?* Washington, D.C.: Center for Advanced Concepts and Technology, Institute for National Strategic Studies, National Defense University, 1995b.

———, *Who Runs What in the Global Information Grid: Ways to Share Local and Global Responsibility*, Santa Monica, Calif.: RAND Corporation, MR-1247-AF, 2000. As of March 13, 2008:
http://www.rand.org/pubs/monograph_reports/MR1247/

———, *Conquest in Cyberspace: National Security and Information Warfare*, New York: Cambridge University Press, 2007.

Libicki, Martin C., Peter Chalk, and Melanie Sisson, *Exploring Terrorist Targeting Preferences*, Santa Monica, Calif.: RAND Corporation, MG-483-DHS, 2007. As of June 27, 2008:
http://www.rand.org/pubs/monographs/MG483/

Lifton, Robert Jay, *Destroying the World to Save It: Aum Shinrikyo, Apocalyptic Violence, and the New Global Terrorism*, New York: Henry Holt and Co., 1999.

Linzer, Dafna, and Thomas E. Ricks, "Anbar Picture Grows Clearer, and Bleaker," *Washington Post*, November 28, 2006, p. A1.

Luttwak, Edward, *Strategy: The Logic of War and Peace*, Cambridge, Mass.: Belknap Press of Harvard University Press, 2001.

Macdonald, Andrew, *The Turner Diaries*, 2nd ed., New York: Barricade Books, 1996.

Mandela, Nelson, "An Ideal for Which I Am Prepared to Die: Nelson Mandela April 20 1964," *Guardian*, April 23, 2007.

McClintock, Cynthia, *Revolutionary Movements in Latin America: El Salvador's FMLN and Peru's Shining Path*, Washington, D.C.: U.S. Institute of Peace Press, 1998.

McClintock, Michael, *The American Connection*, London: Zed Books, 1985.

McConnell, J. Michael, *Annual Threat Assessment of the Director of National Intelligence for the Senate Select Committee on Intelligence*, February 5, 2008.

McCormick, David H., "From Peacekeeping to Peacebuilding: Restructuring Military and Police Institutions in El Salvador," in Michael W. Doyle, Ian Johnstone, and Robert C. Orr, eds., *Keeping the Peace: Multidimensional UN Operations in Cambodia and El Salvador*, New York: Cambridge University Press, 1997, pp. 282–311.

MEMRI—*see* Middle East Media Research Institute.

Michaels, Jim, "Behind Success in Ramadi: An Army Colonel's Gamble," *USA Today*, May 1, 2007a, p. A1.

———, "Tribes Help U.S. Against al-Qaeda: Baghdad-Area Deals Build on Successes in Anbar Province," *USA Today*, June 20, 2007b, p. A1.

Middle East Media Research Institute, *Islamist Sheikh abu Osama al-'Iraqi Denounces al-Qaeda in Iraq for Atrocities Against Sunnis*, Washington, D.C., special dispatch 1340, October 31, 2006.

———, *The Islamic State of Iraq Issues Regulations for Women Drivers*, Washington, D.C., special dispatch 1514, March 23, 2007a.

———, *Continued Clashes in Iraq Between Sunni Jihad Groups and al-Qaeda*, Washington, D.C., special dispatch 1542, April 13, 2007b.

Mine, Douglas Grant, "Guerrillas Attack Affluent Neighborhoods," Associated Press, November 29, 1989.

MIPT—*see* National Memorial Institute for the Prevention of Terrorism.

Mishal, Shaul, and Avraham Sela, *The Palestinian Hamas: Vision, Violence, and Coexistence*, New York: Columbia University Press, 2000.

Montgomery, Tommie Sue, *Revolution in El Salvador: From Civil Strife to Civil Peace*, 2nd ed., Boulder, Colo.: Westview Press, 1995.

Moore, Solomon, and Louise Roug, "Deaths Across Iraq Show It Is a Nation of Many Wars, with U.S. in the Middle," *Los Angeles Times*, October 7, 2006, p. 1.

Morgan, T. Clifton, and Valerie L. Schwebach, "Fools Suffer Gladly: The Use of Economic Sanctions in International Crises," *International Studies Quarterly*, Vol. 41, No. 1, March 1997, pp. 27–50.

Mousseau, Michael, "Market Civilization and Its Clash with Terror," *International Security*, Vol. 27, No. 3, Winter 2002–2003, pp. 5–29.

Muller, Edward N., "Income Inequality, Regime Repressiveness, and Political Violence," *American Sociological Review*, Vol. 50, No. 1, February 1985, pp. 47–61.

Mullins, Mark, "The Political and Legal Response to Aum-Related Violence in Japan: A Review Article," *Japan Christian Review*, Vol. 63, 1997, pp. 37–46.

———, "The Legal and Political Fallout of the 'Aum Affair,'" in Robert Kisala and Mark Mullins, eds., *Religion and Social Crisis in Japan: Understanding Japanese Society Through the Aum Affair*, New York: Palgrave, 2001, pp. 71–72.

Munck, Gerardo L., and Dexter Boniface, "Political Processes and Identity Formation in El Salvador: From Armed Left to Democratic Left," in Ronaldo Munck and Purnaka L. De Silva, eds., *Postmodern Insurgencies: Political Violence, Identity Formation, and Peacemaking in Comparative Perspective*, New York: St. Martin's Press, 2000, pp. 38–53.

Musharraf, Pervez, *In the Line of Fire: A Memoir*, New York: Free Press, 2006.

Naji, Abu Bakr, *The Management of Savagery: The Most Critical Stage Through Which the Umma Will Pass*, Cambridge, Mass.: John M. Olin Institute for Strategic Studies, Harvard University, 2006.

National Commission on Terrorist Attacks upon the United States, *The 9/11 Commission Report: Final Report of the National Commission on Terrorist Attacks upon the United States*, Washington, D.C., 2004.

National Intelligence Council, *Al-Qa'ida, the Organization: A Five-Year Forecast*, Washington, D.C.: Central Intelligence Agency, 2007.

National Memorial Institute for the Prevention of Terrorism, *MIPT Terrorism Knowledge Base: A Comprehensive Databank of Global Terrorist Incidents and Organizations*, Oklahoma City, Okla., ongoing since 2003. As of March 5, 2008: http://www.tkb.org

National Police Agency, *1996 Police White Paper*, Tokyo, 1996.

New York City Police Department, *Threat Analysis: Subject: JFK Airport/Pipeline Plot*, New York, June 2, 2007.

NIO—*see* Great Britain Northern Ireland Office.

NPA—*see* National Police Agency.

NYPD—*see* New York City Police Department.

Office of the Director of National Intelligence, and National Intelligence Council, *The Terrorist Threat to the US Homeland*, Washington, D.C., 2007.

Office of Management and Budget, *The Budget of the United States Government*, Washington, D.C.: Executive Office of the President, Office of Management and Budget, last updated January 28, 2008.

Olson, Kyle B., "Aum Shinrikyo: Once and Future Threat?" *Emerging Infectious Diseases*, Vol. 5, No. 4, July–August 1999, p. 515.

OMB—*see* Office of Management and Budget.

Onishi, Norimitsu, "British Plans to Leave Sierra Leone Prompt Worry," *New York Times*, June 7, 2000, p. A14.

Oppel, Richard A., "Mistrust as Iraqi Troops Encounter New U.S. Allies," *New York Times*, July 16, 2007, p. A1.

Pangi, Robyn, "Consequence Management in the 1995 Sarin Attacks on the Japanese Subway System," *Studies in Conflict and Terrorism*, Vol. 25, No. 6, November–December 2002, pp. 421–448.

Pape, Robert A., "Why Economic Sanctions Do Not Work," *International Security*, Vol. 22, No. 2, Autumn 1997, pp. 90–136.

———, "Why Economic Sanctions Still Do Not Work," *International Security*, Vol. 23, No. 1, Summer 1998, pp. 66–77.

———, "The Strategic Logic of Suicide Terrorism," *American Political Science Review*, Vol. 97, No. 3, August 2003, pp. 343–361.

———, *Dying to Win: The Strategic Logic of Suicide Terrorism*, New York: Random House, 2005.

Parachini, John, "Aum Shinrikyo," in Brian A. Jackson, John C. Baker, Peter Chalk, Kim Cragin, John V. Parachini, and Horacio R. Trujillo, *Aptitude for Destruction*, Vol. 2: *Case Studies of Organizational Learning in Five Terrorist Groups*, Santa Monica, Calif.: RAND Corporation, MG-332-NIJ, 2005, pp. 11–36. As of March 6, 2008: http://www.rand.org/pubs/monographs/MG332/

Pearce, Jenny, "From Civil War to 'Civil Society': Has the End of the Cold War Brought Peace to Central America?" *International Affairs*, Vol. 74, No. 3, July 1998, pp. 587–615.

Perito, Robert, *The American Experience with Police in Peace Operations*, Clementsport, N.S.: Canadian Peacekeeping Press, 2002.

Perlez, Jane, "Briton Criticizes U.S.'s Use of 'War on Terror,'" *New York Times*, April 17, 2007, p. A10.

Perry, Tony, "The Conflict in Iraq: Struggles in Al Anbar," *Los Angeles Times*, December 15, 2006, p. A17.

———, "The Conflict in Iraq: Violence in the Capital," *Los Angeles Times*, January 23, 2007, p. A5.

Peterson, Anna Lisa, *Martyrdom and the Politics of Religion: Progressive Catholicism in El Salvador's Civil War*, Albany, N.Y.: State University of New York Press, 1997.

Pew Research Center, *The Great Divide: How Westerners and Muslims View Each Other*, Washington, D.C., 2006.

Pillar, Paul R., *Negotiating Peace: War Termination as a Bargaining Process*, Princeton, N.J.: Princeton University Press, 1983.

———, *Terrorism and U.S. Foreign Policy*, Washington, D.C.: Brookings Institution Press, 2001.

Posen, Barry, "The Struggle Against Terrorism: Grand Strategy, Strategy, and Tactics," *International Security*, Vol. 26, No. 3, Winter 2001–2002, p. 39–55.

Prendes, Jorge Caceres, "Revolutionary Struggle and Church Commitment: The Case of El Salvador," *Social Compass*, Vol. 30, Nos. 2–3, 1983, pp. 261–298.

Priest, Dana, and Susan Schmidt, "Al Qaeda's Top Primed to Collapse, U.S. Says: Mohammed's Arrest, Data Breed Optimism," *Washington Post*, March 16, 2003, p. A1.

Public Law 93-669, Foreign Assistance Act, 1974.

Public Security Intelligence Agency, *The Review and Prospects of Internal and External Situations*, Tokyo, 2006.

Rabasa, Angel, Cheryl Benard, Lowell H. Schwartz, and Peter Sickle, *Building Moderate Muslim Networks*, Santa Monica, Calif.: RAND Corporation, MG-574-SRF, 2007. As of March 11, 2008:
http://www.rand.org/pubs/monographs/MG574/

Ragin, Charles C., "Comparative Sociology and the Comparative Method," *International Journal of Comparative Sociology*, Vol. 22, Nos. 1–2, March–June 1981, pp. 102–120.

Rapoport, David C., "Fear and Trembling: Terrorism in Three Religious Traditions," *American Political Science Review*, Vol. 78, No. 3, September 1984, pp. 658–677.

———, "Terrorism," in Lester R. Kurtz and Jennifer E. Turpin, eds., *Encyclopedia of Violence, Peace, and Conflict*, San Diego, Calif.: Academic Press, 1999, pp. 497–510.

———, "The Fourth Wave: September 11 in the History of Terrorism," *Current History*, Vol. 100, December 2001, pp. 419–424.

———, "The Four Waves of Modern Terrorism," in Audrey Kurth Cronin and James M. Ludes, eds., *Attacking Terrorism: Elements of a Grand Strategy*, Washington, D.C.: Georgetown University Press, 2004, pp. 46–73.

Reader, Ian, *Religious Violence in Contemporary Japan: The Case of Aum Shinrikyo*, Honolulu: University of Hawai'i Press, 2000.

———, "Spectres and Shadows: Aum Shinrikyo and the Road to Megiddo," *Terrorism and Political Violence*, Volume 14, Number 1, Spring 2002, pp. 145–186.

Reich, Walter, ed., *Origins of Terrorism: Psychologies, Ideologies, Theologies, States of Mind*, Washington, D.C.: Woodrow Wilson Center Press, 1998.

"Religious Scholars Call on Taliban to Abandon Violence," *Pajhwok Afghan News*, July 28, 2005.

Riedel, Bruce, "Al Qaeda Strikes Back," *Foreign Affairs*, Vol. 86, No. 3, May–June 2007, p. 24.

Rosenau, William, "The Kennedy Administration, U.S. Foreign Internal Security Assistance and the Challenge of 'Subterranean War,' 1961–63," *Small Wars and Insurgencies*, Vol. 14, No. 3, Autumn 2003, pp. 65–99.

Ross, Jeffrey Ian, and Ted Robert Gurr, "Why Terrorism Subsides: A Comparative Study of Canada and the United States," *Comparative Politics*, Vol. 21, No. 4, July 1989, pp. 405–426.

Röttger, Maike, "Polizei Zeigt Video der Kölner Bahnbomber," *Hambuger Abendblatt*, August 19, 2006.

Rubin, Barnett R., *The Fragmentation of Afghanistan: State Formation and Collapse in the International System*, 2nd ed., New Haven, Conn.: Yale University Press, 2002.

Rubin, Trudy, "New Iraq Tribal Alliances Fighting al-Qaeda," *Philadelphia Inquirer*, June 22, 2007, p. A19.

Rumsfeld, Donald H., secretary of defense, "Global War on Terrorism," memorandum to General Richard B. Myers, Paul Wolfowitz, General Peter Pace, and Douglas J. Feith, October 16, 2003.

Sageman, Marc, *Understanding Terror Networks*, Philadelphia, Pa.: University of Pennsylvania Press, 2004.

Saleh, Amrullah, *Strategy of Insurgents and Terrorists in Afghanistan*, Kabul: National Directorate for Security, 2006.

Scarborough, Rowan, "Petraeus Adviser: Al Qaeda Weakened by Troop Surge," *Washington Examiner*, August 15, 2007, p. 15.

Schachter, Jonathan, *The Eye of the Believer: Psychological Influences on Counter-Terrorism Policy-Making*, Santa Monica, Calif.: RAND Corporation, RGSD-166, 2002. As of March 6, 2008:
http://www.rand.org/pubs/rgs_dissertations/RGSD166/

Schelling, Thomas, "What Purposes Can 'International Terrorism' Serve?" in R. G. Frey and Christopher W. Morris, eds., *Violence, Terrorism, and Justice*, Cambridge and New York: Cambridge University Press, 1991, pp. 18–32.

Scheuer, Michael, *Imperial Hubris: Why the West Is Losing the War on Terror*, Washington, D.C.: Brassey's, 2004.

Schroen, Gary C., *First In: An Insider's Account of How the CIA Spearheaded the War on Terror in Afghanistan*, New York: Presidio Press/Ballantine Books, 2005.

Schwarz, Benjamin, *American Counterinsurgency Doctrine and El Salvador: The Frustrations of Reform and the Illusions of Nation Building*, Santa Monica, Calif.: RAND Corporation, R-4042-USDP, 1991. As of March 7, 2008:
http://www.rand.org/pubs/reports/R4042/

Sheehan, Michael A., "Diplomacy," in Audrey Kurth Cronin and James M. Ludes, eds., *Attacking Terrorism: Elements of a Grand Strategy*, Washington, D.C.: Georgetown University Press, 2004, pp. 104–105.

Sims, Calvin, "Poison Gas Group in Japan Distances Itself from Guru," *New York Times*, January 19, 2000, p. A6.

Skocpol, Theda, and Margaret Somers, "The Uses of Comparative History in Macrosocial Inquiry," *Comparative Studies in Society and History*, Vol. 22, No. 2, April 1980, pp. 174–197.

Smith, Major Neil, and Colonel Sean MacFarland, "Anbar Awakens: The Tipping Point," *Military Review*, March–Aprril 2008.

Soto, Alvaro de, representative of the secretary-general of the United Nations, "Note Verbale Dated 90/08/14 from the Charge D'Affaires A.I. of the Permanent Mission of El Salvador to the United Nations Addressed to the Secretary-General," New York: United Nations, A/44/971-S-21541, August 16, 1990.

Sprinzak, Ehud, "Rational Fanatics," *Foreign Policy*, No. 120, September–October 2000, pp. 66–73.

Stanley, William Deane, *The Protection Racket State: Elite Politics, Military Extortion, and Civil War in El Salvador*, Philadelphia, Pa.: Temple University Press, 1996.

Stedman, Stephen John, "Negotiation and Mediation in Internal Conflict," in Michael E. Brown, ed., *The International Dimensions of Internal Conflict*, Cambridge, Mass.: MIT Press, 1996, pp. 341–376.

Stern, Jessica, *Terror in the Name of God: Why Religious Militants Kill*, New York: HarperCollins Publishers, 2003.

Stoltenberg, Jochim, "Jochim Stoltenberg zu den Fahndungsergebnissen über die Bombenleger von Dortmund und Koblenz," *Berliner Morgenpost*, August 19, 2006.

Straziuso, Jason, "2007 Was a Year of Record Violence in Afghanistan, but U.S. Says Things Are Looking Up," Associated Press, January 1, 2008.

"Taliban Claim Killing of Pro-Government Religious Scholars in Helmand," *Afghan Islamic Press*, July 13, 2005.

Taylor, Maxwell D., "Address at International Police Academy Graduation," press release, U.S. Agency for International Development, December 17, 1965.

"Terrorverdacht Wird Konkreter," *Süddeutsche Zeitung*, August 12, 2006.

Tilly, Charles, "Means and Ends of Comparison in Macrosociology," in Fredrik Engelstad, ed., *Comparative Social Research*, Vol. 16: *Methodological Issues in Comparative Social Science*, London: JAI, 1997, pp. 43–53.

Tilly, Charles, and Gabriel Ardant, *The Formation of National States in Western Europe*, Princeton, N.J.: Princeton University Press, 1975.

Torres-Rivas, Edelberto, "Insurrection and Civil War in El Salvador," in Michael W. Doyle, Ian Johnstone, and Robert C. Orr, eds., *Keeping the Peace: Multidimensional UN Operations in Cambodia and El Salvador*, New York: Cambridge University Press, 1997, pp. 209–226.

Trager, Robert, and Dessislava P. Zagorcheva, "Deterring Terrorism: It Can Be Done," *International Security*, Vol. 30, No. 3, Winter 2005–2006, pp. 87–123.

Trinquier, Roger, *Modern Warfare: A French View of Counterinsurgency*, New York: Praeger, 1964.

UN/DPI—*see* United Nations Department of Public Information.

UNSG—*see* United Nations Secretary-General.

United Nations Department of Peacekeeping Operations, Cartographic Section, "Iraq," map 3835, rev. 4, January 2004.

United Nations Department of Public Information, *The United Nations and El Salvador, 1990–1995*, New York, 1995a.

———, *The United Nations and Mozambique, 1992–1995*, New York, 1995b.

United Nations Secretary-General, *Report of the Secretary-General on the United Nations Observer Mission in El Salvador (ONUSAL)*, New York: United Nations, S/25006, December 12, 1992.

———, *Report of the Secretary-General on the United Nations Observer Mission in El Salvador*, New York: United Nations, S/25812/Add.3, May 25, 1993.

———, *Report of the Secretary-General on the United Nations Observer Mission in El Salvador*, New York: United Nations, S/1994/179, February 16, 1994a.

———, *Report of the Secretary-General on the United Nations Observer Mission in El Salvador*, New York: United Nations, S/1994/304, March 16, 1994b.

———, *Report of the Secretary-General on the United Nations Observer Mission in El Salvador*, New York: United Nations, S/1994/375, March 31, 1994c.

United States v Mokhtar Haouari and Abdelghani Meskini, 2001 U.S. Dist. LEXIS 6566, S.D.N.Y., May 18, 2001.

United States v Russell Defreitas, Kareem Ibrahim, Abdul Kadir, and Abdel Nur, indictment, E.D.N.Y., filed June 28, 2007.

U.S. Attorney's Office, Eastern District of New York, "Four Individuals Charged in Plot to Bomb John F. Kennedy International Airport," press release, Brooklyn, N.Y., June 2, 2007.

U.S. Department of the Army, and U.S. Marine Corps, *Counterinsurgency*, Washington, D.C., field manual 3-24, 2006.

U.S. Department of Defense, *Quadrennial Defense Review Report*, Washington, D.C., 2006.

U.S. Department of Justice, U.S. Attorney's Office, Eastern District of New York, "Four Individuals Charged in Plot to Bomb John F. Kennedy International Airport," press release, Washington, D.C., June 2, 2007.

U.S. Department of State, *Communist Interference in El Salvador: Documents Demonstrating Communist Support of the Salvadoran Insurgency*, Washington, D.C., 1981.

———, Office of the Secretary of State, Office of the Coordinator for Counterterrorism, *Patterns of Global Terrorism, 1995*, Washington, D.C., 1996.

———, Office of the Coordinator for Counterterrorism, *Country Reports on Terrorism 2005*, Washington, D.C., 2006.

———, Office of the Coordinator for Counterterrorism, "NCTC Observations Related to Terrorist Incidents Statistical Material," *Country Reports on Terrorism 2006*, Washington, D.C., 2007.

U.S. Institute of Peace, *How Terrorism Ends*, Washington, D.C., 1999.

U.S. Joint Chiefs of Staff, *Department of Defense Dictionary of Military and Associated Terms*, Washington, D.C., joint publication 1-02, ongoing since 1972.

———, *Information Operations*, Washington, D.C., February 13, 2006.

U.S. Senate Committee on Foreign Relations, and U.S. House of Representatives Committee on International Relations, *Legislation on Foreign Relations Through 2000*, Washington, D.C.: U.S. Government Printing Office, 2001.

U.S. Senate Committee on Governmental Affairs Permanent Subcommittee on Investigations, *Staff Statement, U.S. Senate Permanent Subcommittee on Investigations (Minority Staff): Hearings on Global Proliferation of Weapons of Mass Destruction: A Case Study on the Aum Shinrikyo*, Washington, D.C., 1995.

USIP—*see* U.S. Institute of Peace.

USJCS—*see* U.S. Joint Chiefs of Staff.

Van Evera, Stephen, *Guide to Methods for Students of Political Science*, Ithaca, N.Y.: Cornell University Press, 1997.

Vásquez Perdomo, María Eugenia, *My Life as a Colombian Revolutionary: Reflections of a Former Guerrillera*, Philadelphia, Pa.: Temple University Press, 2005.

Vilas, Carlos María, *Between Earthquakes and Volcanoes: Market, State, and the Revolutions in Central America*, New York: Monthly Review Press, 1995.

Voice of Jihad, "Imarat Islami of Afghanistan," undated Web page. As of April 14, 2008:
http://www.alemarah.8rf.com/

"Waiting for al-Qaeda's Next Bomb: MI5 and al-Qaeda," *Economist*, Vol. 383, No. 8527, May 5, 2007, pp. 29–31.

Walter, Barbara F., *Committing to Peace: The Successful Settlement of Civil Wars*, Princeton, N.J.: Princeton University Press, 2002.

Weber, Max, "Politics as a Vocation," in Max Weber, Hans Heinrich Gerth, and C. Wright Mills, eds., *From Max Weber: Essays in Sociology*, New York: Oxford University Press, 1958, pp. 77–128.

Wehrfritz, George, "Crushing the Cult of Doom," *Newsweek*, November 22, 1999, p. 44.

Weinberg, Leonard, and William Lee Eubank, *The Rise and Fall of Italian Terrorism*, Boulder, Colo.: Westview Press, 1987.

West, Bing, and Owen West, "Iraq's Real 'Civil War,'" *Wall Street Journal*, April 5, 2007, p. A13.

Whitfield, Teresa, *Paying the Price: Ignacio Ellacuría and the Murdered Jesuits of El Salvador*, Philadelphia, Pa.: Temple University Press, 1994.

Wilkins, Timothy A., "The El Salvador Peace Accords: Using International and Domestic Law Norms to Build Peace," in Michael W. Doyle, Ian Johnstone, and Robert C. Orr, eds., *Keeping the Peace: Multidimensional UN Operations in Cambodia and El Salvador*, New York: Cambridge University Press, 1997, pp. 275–277.

Wilkinson, Paul, *Terrorism and the Liberal State*, 2nd ed., Basingstoke, UK: Macmillan, 1986.

Wong, Edward, and Khalid al-Ansary, "Iraqi Sheiks Assail Cleric for Backing Qaeda," *New York Times*, November 19, 2006, p. 22.

Wood, Elisabeth Jean, *Forging Democracy from Below: Insurgent Transitions in South Africa and El Salvador*, Cambridge and New York: Cambridge University Press, 2000.

Woodward, Bob, *Bush at War*, New York: Simon and Schuster, 2002.

Woodwell, Douglas, "The 'Troubles of Northern Ireland': Civil Conflict in an Economically Well-Developed State," in Paul Collier and Nicholas Sambanis, eds., *Understanding Civil War: Evidence and Analysis*, Vol. 2: *Europe, Central Asia, and Other Regions*, Washington, D.C.: World Bank, 2005, pp. 161–190.

World Bank, *World Development Indicators 2007*, Washington, D.C., 2007.

Wright, Lawrence, *The Looming Tower: Al Qaeda and the Road to 9/11*, New York: Knopf, 2006.

Wrobel, Paulo S., and Guilherme Theophilo Gaspar de Oliverra, *Managing Arms in Peace Processes: Nicaragua and El Salvador*, New York: United Nations, 1997.

Yousafzai, Sami, and Ron Moreau, "Unholy Allies," *Newsweek*, September 26, 2005, pp. 40–42.

Zartman, I. William, "The Unfinished Agenda: Negotiating Internal Conflicts," in Roy E. Licklider, ed., *Stopping the Killing: How Civil Wars End*, New York: New York University Press, 1993, pp. 25–26.

Index